# Turning the World Inside Out

## AND 174 OTHER SIMPLE PHYSICS DEMONSTRATIONS

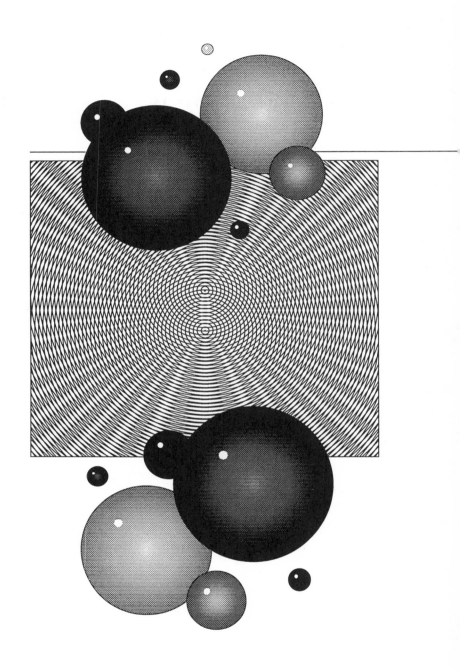

# Turning the World Inside Out

AND 174 OTHER SIMPLE PHYSICS DEMONSTRATIONS

Robert Ehrlich

**Princeton University Press** • Princeton, New Jersey

Copyright © 1990 by
Princeton University
Press
Published by
Princeton University
Press, 41 William Street,
Princeton, New Jersey
08540
In the United Kingdom:
Princeton University
Press, Oxford
All Rights Reserved

Library of Congress Cataloging-in-Publication Data

Ehrlich, Robert, 1938–
Turning the world inside out and 174 other simple physics
demonstrations / Robert Ehrlich.
p.   cm.
Includes bibliographical references.
ISBN 0-691-08534-X (alk. paper)
ISBN 0-691-02395-6 (pbk. : alk. paper)
1. Physics—Experiments.   I. Title.
QC33.E54     1990
530'.076—dc20          89-36976 CIP

This book has been composed in Adobe Times Roman and Helvetica
typefaces on a Varityper VT-600 laser printer, using Bestinfo's
PageWright composition software

Princeton University Press books are printed on acid-free paper, and meet
the guidelines for permanence and durability of the Committee on
Production Guidelines for Book Longevity of the
Council on Library Resources

Printed in the United States of America by
Princeton University Press, Princeton, New Jersey

10   9   8   7   6   5   4   3   2   1

(Pbk.)
10   9   8   7   6   5   4   3   2   1

Cover illustration: "Mercury$^2$," by Hubert C. Delany. Copyright © 1988
by Hubert C. Delany. All Rights Reserved. Photograph by Jon Chomitz.
The image and the software used to create this illustration were developed
with support from the Massachusetts Institute of Technology Media Laboratory and from Thinking Machines Corporation. The image was produced at $1024 \times 1024$ pixels with five orders of reflection in six minutes
using a Connection Machine II. The algorithm used to produce the picture
is called "Induced-Octree Ray Tracing" and is described in the paper
"Ray Tracing on a Connection Machine," by Hubert C. Delany, *1988
ACM International Conference on Supercomputing*, July 4–8, St. Malo,
France, pp. 659–67.

All demonstrations in this book have been tested by the author and are
safe if performed as directed. Demonstrations marked as potentially hazardous should be performed only by or under the supervision of qualified
laboratory instructors. Princeton University Press and the author assume
no liability for injuries or damages caused by performing or modifying
these demonstrations.

*This book is dedicated*
*to the memory of my mother,*

**Frances Ehrlich**

# Contents

Contents

# List of Plates

List of Plates

# Foreword

Here is a treasure chest of ideas that demonstrate aspects of physics and physical science in simple, playful ways. Whether you are a teacher, student, or just someone who enjoys science, dipping into this collection is much like opening a holiday gift and discovering a marvelous little toy that then holds your attention by some curious performance. Here are, in effect, nearly two hundred such toys.

This book precisely reflects the way science education should be, especially at the introductory level. The primary reason scientists become scientists is that science is a thoughtful form of play. It is fun to see an arrangement of common items behave in a surprising way and to try to figure out why it acts that way. We learn science because it reveals the otherwise hidden simplicity of how the world works, and often that revelation is so surprising that we might even laugh.

That is the way science should be taught, but too often it is demonstrated with large, noisy, foreboding apparatus, things that students or other onlookers find alien and removed from everyday life. Sometimes the element of strangeness is so strong that even a simple principle may be obscured by it. Such apparatus may be impressive but may not be educational, because they and the principles they demonstrate may be quickly forgotten. The demonstrations that are long remembered (and retold to others) are the simple ones that are fashioned from everyday objects.

Ehrlich has gathered together here simple demonstrations, each tested and tagged with a concise and accessible explanation. If you are a teacher, you will find them especially valuable because they have been chosen for a minimum of expense and setup time. If you are not a teacher, you will find them just as valuable because they open up the world of science without needing costly or intimidating equipment.

Once read, this book serves another purpose—you will begin to think like Ehrlich and discover other, similarly simple demonstrations that can be made from common objects.

This book begins the fun and you then will continue it, sometimes laughing along the way when you can coax another hidden simplicity of the world out into the open.

JEARL WALKER
Department of Physics
Cleveland State University
author of "The Amateur Scientist," which appears monthly in *Scientific American*, and *The Flying Circus of Physics with Answers*

# Acknowledgments

I am grateful to the many people who made helpful suggestions during the preparation of this book, especially the following people who made comments on an early draft: James Bond, Robert W. Ellsworth, Michael Randy Gabel, Mary Lynn Hutchison, Robert Karlson, Joan Sulivan Kowalski, Malgorzata Z. Pfabe, Patricia Rourke, Robert J. Salewski, George Smith, and James Trefil. I am also grateful to the students in my classes for their feedback on a preliminary version of the book. I thank Dr. James A. Yorke for his permission to use Plate C.14b, which he kindly provided. I especially wish to thank my son David for the photographs used in the book, my son Gary for the drawings used, and my wife Elaine for introducing me to the wonderful world of yard sales—an endless source of inexpensive physics demonstration equipment. Needless to say, any errors in the book are entirely my responsibility.

The cover image, "Mercury$^2$," has been supplied by Hubert C. Delany, copyright 1988 by Hubert C. Delany, all rights reserved. The image and the software used to create the illustration were developed with support from the MIT Media Laboratory and Thinking Machines Corporation. This computer-generated image (photograph by Jon Chomitz) was produced at $1024 \times 1024$ pixels with 5 orders of reflection in 6 minutes using a Connection Machine II. The algorithm used to produce the picture is called "Induced-Octree Ray Tracing," and is described in the paper "Ray Tracing on a Connection Machine," Hubert C. Delany, *1988 ACM International Conference on Supercomputing*, July 4–8, St. Malo, France, pp. 659–67.

# Introduction

Anyone curious about the physical universe cannot help but be intrigued by simple demonstrations of physical principles at work. Albert Einstein wrote that as a small boy he first became aware of a hidden order in the universe by witnessing the behavior of a compass. We may speculate that his lifelong search for order, even though couched in esoteric mathematics, had its roots in simple observations and in simple "thought experiments." Even those of us lacking Einstein's deep insight and mathematical abilities can still find our understanding of the physical universe deepened, or our curiosity aroused, by witnessing a range of simple demonstrations.

This book is a collection of physics demonstrations for teachers, students, and laypersons who have a curiosity about the physical world. Many are suitable for use in a variety of educational settings, ranging from middle-school physical science classes to university-level physics courses, but because of their simplicity most can be performed by students outside of class—indeed, by anyone.

As a university physics teacher I have noticed that most of my colleagues love to watch a good demonstration, but rarely do demonstrations in their own classes. The problem seems to be that, except for those fortunate enough to teach at a university with a staffed demonstration facility, setting up demonstrations for classes is too much effort. There is an enormous amount of work involved in designing, constructing, physically locating, and setting up a demonstration that works reliably. Accordingly, I have written this book with the intention of compiling a list of demonstrations whose simplicity and convenience will, I hope, stimulate increased use of demonstrations, both in and out of class, and perhaps even turn the attention of some future Einstein toward the mysteries of the physical world.

The primary characteristic shared by most of the demonstrations in this book is transparency. The experiments are transparent in two senses: (1) they reveal the underlying physics in as simple a way as possible, and (2) they are literally transparent—many can be done on an overhead projector, particularly important for classroom use. Very few of the demonstrations, however, require the use of an overhead projector, so general readers or physics students wishing to perform a variety of physics demonstrations on their own can easily disregard references to the overhead projector and the use of transparent materials. In choosing demonstrations to include in the book, I have also applied a number of additional criteria:

*Low cost*   Wherever possible, inexpensive items are used (under $20.00 in almost all cases, and usually much less).

*Simplicity of design*   In all cases, designs have been perfected to make use of readily obtainable materials, and a minimum of skill is required to implement the design. A good example of this philosophy is the design recommended for a homemade ripple tank (demonstration Q.3) that gives results as good as any commercially available ripple tank.

*Feasibility*   In assembling this collection, I have avoided dramatic but temperamental demonstrations, which can be a great source of frustration (and a deterrent to doing demonstrations!). All the demonstrations included actually work as advertised. Please contact me if you find this not to be the case.

*Safety*   Wherever possible, unsafe materials and high voltages have been avoided. A number of somewhat hazardous and dramatic demonstrations have been omitted on this basis, and safety tips have been listed for all demonstrations that present some degree of hazard.

*Compactness*   Having personally been frustrated by demonstrations that were very large and not portable (and unusable if your class is not in a lecture hall next to a demonstration facility), I have learned the virtue of compact items that can be easily carried to class.

*Setup time*   All the demonstrations in this book require almost no time to set up, and the large majority can be performed in under one minute—important when one is pressed for time.

*Quantitative results*   Although many demonstrations present a phenomenon only qualitatively, many others permit measurements and calculations to be made that allow a quantitative check on various physical principles. In discussing such demonstrations, I have focused primarily on their quantitative aspect in order to show how far the demonstrations could be carried. It should be clear, however, that all demonstrations capable of quantitative results can also be done qualitatively. For example, instead of directly verifying the exponential decay law for a capacitor discharging through a voltmeter by measuring the voltage at a series of times, you could just convey the sense of an exponential decay by showing how the voltmeter needle swings toward zero with a gradually decreasing speed and yet never quite gets there.

Because the demonstrations are simple, compact, and inexpensive, it is feasible for instructors to have their own dedicated collection stored in a cabinet in their office. Moreover, the same properties that make the demonstrations suitable for use in lecture classes should also make them suitable for student use, either in or out of class. The demonstrations listed all include very specific details on construction where needed, and they also include a complete list of pitfalls to be avoided in conducting the demonstration. Attention to such details often means the difference between a nice idea that fails to work in practice and a dramatic reproducible demonstration. Spectacular "gee-whiz" demonstrations whose point is obscure have been avoided, however.

Many demonstrations include some brief explanation of the theory, although it is assumed that the reader already has some familiarity with elementary physics, including mechanics, heat, electricity, magnetism, optics, and waves. Although the majority of demonstrations are not completely new, many are, and others represent new twists on old ideas. For many well-known demonstrations, such as the singing wine glass, the tornado in a bottle, and the "whirl-a-tube," the correct explanation of the physics involved is *not* well-known, and is included here.

The 175 demonstrations have been grouped in the text according to the various areas of physics. Some readers may be interested in features of a demonstration other than the area of physics it illustrates. The following list of non-physics categories gives the number of demonstrations that fall into each category in parentheses.

**Non-Physics Categories of Demonstrations**
Surprising result (86)
New demonstration (66)
Cost: near zero (73)
Cost: between "near zero" and $2 (55)
Cost: between $2 and $12 (31)
Quantitative component (67)

Overhead projector optional (57)
Overhead projector required (8)
Some construction required (32)
Some degree of hazard (9)
Analogy (simulation) (6)
Duration: more than one minute (37)

The following six demonstrations make use of analogy (simulation): the three demonstrations in section B (Gravity and Curved Space-Time), demonstration J.2 (Rolled-up transparency for transverse waves), J.4 (Rolled-up transparency for longitudinal waves), and Q.1 (Superposition of circle transparencies).

Readers interested in demonstrations that have a particular property or combination of properties—say, inexpensive demonstrations giving surprising results, suitable for an overhead projector, and doable in under one minute—can easily pick them out in the comprehensive list of demonstrations on the following pages. Some of the categories are highly subjective, none more so than "new" and "surprising." Some well-known demonstrations have been labelled "new" if they contain significant new elements, which means either modified ways of doing the demonstration, or new methods of analyzing or presenting the results. Nevertheless, it is likely that some of these modifications are well known to some readers, and their "newness" merely reflects the author's lack of awareness. The term "surprising" has been applied to demonstrations that would probably surprise the average layperson or physics student, as well as those demonstrations whose theoretical explanation would probably surprise most physics teachers who may have seen the demonstration many times.

In estimating the cost of a demonstration, all items needed are included, even if they are likely to be already in your possession. An excellent source of inexpensive items useful for physics demonstrations are flea markets and private yard sales, also known as garage, tag, or rummage sales in some areas. In addition, many useful items can be obtained in hardware, toy, electronics, craft, and sporting-goods stores. In many of the demonstrations I have identified by name particular store chains or companies. These supplier identifications are for your convenience; they should not be taken to be commercial endorsements. No consideration, financial or otherwise, has been received from companies mentioned in the book. (See Appendix 2 for a list of companies with addresses.)

# Complete List of Demonstrations

## Key

### A. Accelerated Motion and the Acceleration of Gravity

| ! | N | Cost | Q | O/R | C | + | * | Demonstration |
|---|---|------|---|-----|---|---|---|---------------|
|   |   | ¢    |   |     |   | + |   | 1. Dropping balls of different sizes |
| ! | N | $    |   | O   | C |   |   | 2. "Monkey and hunter" demonstration on an incline |
|   | N | ¢    | Q | O   |   |   |   | 3. Rolling balls down an inclined ruler |
|   | N |      | Q | O   | C |   |   | 4. Rolling balls on a vibrating plate |
|   | N |      |   |     |   |   |   | 5. Dropping an accelerometer from various heights |

### B. Gravity and Curved Space-Time

| ! | N | Cost | Q | O/R | C | + | * | Demonstration |
|---|---|------|---|-----|---|---|---|---------------|
|   | N | $    |   | O   |   |   |   | 1. Rolling balls on a stretched membrane |
|   |   | ¢    |   | O   |   |   |   | 2. Simulation of the gravitational deflection of light |
|   |   | ¢    |   | O   |   |   |   | 3. Acceleration as an effect of curved space-time |

### C. Newton's Laws

| ! | N | Cost | Q | O/R | C | + | * | Demonstration |
|---|---|------|---|-----|---|---|---|---------------|
|   |   | ¢    |   | O   |   |   |   | 1. Shooting a penny out from under a stack of pennies |
| ! |   | ¢    |   |     |   |   |   | 2. Catching a row of pennies on your arm |
|   |   | ¢    | Q |     |   |   |   | 3. Chain with a suspended weight |
|   | N | $    | Q | O   | C | + |   | 4. Force table |
| ! | N | $    |   |     |   |   |   | 5. Swinging two balls into a block |
|   |   |      | Q |     |   |   |   | 6. Accelerating a scale with a suspended weight |
|   |   | ¢    |   |     |   |   |   | 7. Pulling a thread attached to a hanging weight |
| ! |   | $    |   |     |   |   | * | 8. "Vampire killer" |
| ! |   | ¢    |   |     | C |   |   | 9. Ping-Pong-ball buoy anchored to a weight in a jar |

## List of Demonstrations

| ! = SURPRISING RESULT |
| N = NEW DEMONSTRATION |
| ¢ = COST: NEAR ZERO |
| $ = COST: BETWEEN "NEAR ZERO" AND $2 |
| $$ = COST: BETWEEN $2 AND $12 |
| Q = QUANTITATIVE COMPONENT |
| O = OVERHEAD PROJECTOR OPTIONAL |
| R = OVERHEAD PROJECTOR REQUIRED |
| C = SOME CONSTRUCTION REQUIRED |
| + = DURATION: MORE THAN ONE MINUTE |
| * IN MARGIN = SOME DEGREE OF HAZARD |

| ! | N | Cost | Q | O | C | + | Demonstration |
|---|---|------|---|---|---|---|---------------|
| ! |   | ¢ |   |   |   |   | **10.** Throwing eggs at a sheet |
|   |   | ¢ |   |   |   |   | **11.** Water rocket, rocket balloon, and balloon-powered helicopter |
|   |   |   |   |   |   |   | **12.** Pulling two scales connected together |
| ! |   | ¢ |   |   |   |   | **13.** Recoil force in a bent straw |
| ! |   | ¢ |   | O | C |   | **14.** Chaotic motion of a pendulum |

**D. Center of Mass, Stability, and Friction**

| ! | N | Cost | Q | O | C | + | Demonstration |
|---|---|------|---|---|---|---|---------------|
| ! |   | ¢ | Q |   |   |   | **1.** Stacking metersticks |
|   | N | ¢ | Q |   |   |   | **2.** Stability of a floating object |
|   |   | $ | Q |   |   |   | **3.** Pulling a sliding brick |
|   | N | ¢ | Q |   |   | + | **4.** Launching a sliding block with known velocity |
|   | N | ¢ | Q |   | C |   | **5.** No tipping allowed |
| ! |   | ¢ |   |   |   | + | **6.** Air resistance as a form of friction |
| ! | N | $ | Q |   | C |   | **7.** Weighted Styrofoam sphere and disk on an incline |
|   | N | ¢ | Q | O |   |   | **8.** Pennies balanced on a ruler |
| ! |   | ¢ | Q |   |   |   | **9.** Moving two fingers under a meterstick |

**E. Energy and Linear Momentum Conservation**

| ! | N | Cost | Q | O | C | + | Demonstration |
|---|---|------|---|---|---|---|---------------|
|   | N | $$ | Q | O |   |   | **1.** Inelastic collisions between two balls |
| ! | N | ¢ | Q |   |   | + | **2.** Rolling disks, hoops, and spheres down an incline |
| ! |   | $ |   |   | C |   | **3.** Rolling a wheel with axle down a slight incline |
|   | N |   | Q | O | C |   | **4.** Collisions on a vibrating plate |
|   | N | $$ | Q | O |   |   | **5.** Loop-the-loop on an incline |
|   |   | ¢ |   | O |   |   | **6.** Colliding balls on a grooved ruler |
|   | N | $ | Q | O |   |   | **7.** Momentum conservation using an embroidery hoop |
| ! |   | ¢ | Q |   |   |   | **8.** Dropping a small ball on top of a big ball |
|   | N | $ |   | O |   |   | **9.** One-dimensional double-well potential |
| ! |   | ¢ |   |   |   |   | **10.** Energy storage in a rubber hemisphere |
| ! | N | $ |   | O | C |   | **11.** Path of least time |

# List of Demonstrations

! = SURPRISING RESULT
N = NEW DEMONSTRATION
¢ = COST: NEAR ZERO
$ = COST: BETWEEN "NEAR ZERO" AND $2
$$ = COST: BETWEEN $2 AND $12
Q = QUANTITATIVE COMPONENT
O = OVERHEAD PROJECTOR OPTIONAL
R = OVERHEAD PROJECTOR REQUIRED
C = SOME CONSTRUCTION REQUIRED
+ = DURATION: MORE THAN ONE MINUTE
* IN MARGIN = SOME DEGREE OF HAZARD

## List of Demonstrations

| ! = SURPRISING RESULT |
| :--- |
| N = NEW DEMONSTRATION |
| ¢ = COST: NEAR ZERO |
| $ = COST: BETWEEN "NEAR ZERO" AND $2 |
| $$ = COST: BETWEEN $2 AND $12 |
| Q = QUANTITATIVE COMPONENT |
| O = OVERHEAD PROJECTOR OPTIONAL |
| R = OVERHEAD PROJECTOR REQUIRED |
| C = SOME CONSTRUCTION REQUIRED |
| + = DURATION: MORE THAN ONE MINUTE |
| * IN MARGIN = SOME DEGREE OF HAZARD |

### H. Fluids

**Pressure and Buoyancy:**

| ! | N | Cost | Q | O/R | C | + | # | Demonstration |
|---|---|------|---|-----|---|---|---|---------------|
| ! |   | ¢ | Q |   | C |   | 1. | Cartesian diver |
|   | N | ¢ | Q |   | C |   | 2. | Sinking a floating block |
|   |   | ¢ | Q |   |   | + | 3. | Three holes in a water-filled bottle |
|   | N | ¢ | Q |   |   | + | 4. | Speed variation of water flowing out of a hole in a can |
|   | N | ¢ | Q |   |   | + | 5. | Squirting a jet out of a water-filled plastic bottle |
|   |   | $$ | Q |   |   |   | 6. | Pulling two plungers apart |
| ! |   | ¢ |   |   |   |   | 7. | Card under a water-filled bottle |
|   |   | $ | Q |   |   | + | 8. | Hanging a weight from a helium-filled balloon |
|   |   | ¢ | Q |   |   |   | 9. | Lowering a weight into a liquid |

**Bernoulli's Principle:**

| ! | N | Cost | Q | O/R | C | + | # | Demonstration |
|---|---|------|---|-----|---|---|---|---------------|
| ! |   | ¢ |   |   |   |   | 10. | Ping-Pong ball near a water stream |
| ! |   | ¢ |   |   |   |   | 11. | Ping-Pong ball in an inverted funnel |
| ! |   | ¢ | Q |   |   |   | 12. | Blowing a quarter into a cup |
| ! | N | $$ |   |   |   |   | 13. | Fan below two vertical pieces of paper |
| ! |   | ¢ |   | O |   |   | 14. | Vortex rings |
| ! |   | ¢ | Q |   |   |   | 15. | Straw atomizer |

**Surface Tension:**

| ! | N | Cost | Q | O/R | C | + | # | Demonstration |
|---|---|------|---|-----|---|---|---|---------------|
| ! |   | ¢ |   | O |   | + | 16. | Dropping pennies into a water-filled cup |
| ! |   | ¢ |   | O |   |   | 17. | Water bridge |
| ! |   | ¢ |   | O |   |   | 18. | Pepper on a water surface |

### I. Heat, Thermodynamics, and Kinetic Theory

| ! | N | Cost | Q | O/R | C | + | # | Demonstration |
|---|---|------|---|-----|---|---|---|---------------|
|   | N | $$ | Q |   |   | + | 1. | Measuring specific heat by the method of mixtures |
| ! |   | ¢ |   |   | C |   | 2. | Hole in a box |
| ! |   | $ | Q | O |   |   | 3. | Bimetallic jumping disks |
|   | N | $ | Q | O | C | + | 4. | Molecular motion simulation |

## List of Demonstrations

! = SURPRISING RESULT
N = NEW DEMONSTRATION
¢ = COST: NEAR ZERO
$ = COST: BETWEEN "NEAR ZERO" AND $2
$$ = COST: BETWEEN $2 AND $12
Q = QUANTITATIVE COMPONENT
O = OVERHEAD PROJECTOR OPTIONAL
R = OVERHEAD PROJECTOR REQUIRED
C = SOME CONSTRUCTION REQUIRED
+ = DURATION: MORE THAN ONE MINUTE
* IN MARGIN = SOME DEGREE OF HAZARD

| ! | N | Cost | Q | O/R | C | + | * | Demonstration |
|---|---|------|---|-----|---|---|---|---------------|
|   |   | $$   |   |     |   |   |   | **5.** Radiometer |
| ! | N | ¢    |   |     |   |   | * | **6.** Burning paper and Styrofoam cups |
|   | N | $$   |   |     | C |   |   | **7.** Heat conductivity demonstration using thermostrips |
|   |   | $$   | Q |     |   | + |   | **8.** Measuring thermal conductivity from rate of heat loss |
|   | N | ¢    | Q |     |   | + |   | **9.** Rate of cooling |
|   |   | $$   | Q |     |   | + |   | **10.** Mechanical equivalent of heat |
| ! | N | $    |   | O   |   |   |   | **11.** Apparent violation of the entropy law |

### J. Waves
### (see section Q for interference and diffraction)

**Traveling Waves:**

| ! | N | Cost | Q | O/R | C | + | Demonstration |
|---|---|------|---|-----|---|---|---------------|
|   |   | $$   | Q |     |   |   | **1.** Waves along a spring or rubber hose |
|   | N | $    |   | O   |   |   | **2.** Rolled-up transparency for transverse waves |
|   |   | $$   | Q |     |   | + | **3.** One-dimensional water wave in a narrow trough |

**Standing Waves:**

| ! | N | Cost | Q | O/R | C | Demonstration |
|---|---|------|---|-----|---|---------------|
|   | N | $    |   | O   |   | **4.** Rolled-up transparency for longitudinal waves |
| ! |   | $    | Q | O   | C | **5.** Sound waves in a variable-length tube |
| ! |   | ¢    |   |     |   | **6.** Snipping a straw as you toot |
| ! |   | $    |   |     |   | **7.** Harmonic frequencies in a corrugated tube |
|   |   | $    |   | O   |   | **8.** Standing waves in a coiled spring |
|   | N | $    |   | O   |   | **9.** Longitudinal waves around a circle |
| ! |   | ¢    |   |     |   | **10.** Rubbing the edge of a wine glass |
| ! |   | $    |   |     |   | **11.** Longitudinal waves in a rod |
|   |   | $    |   |     |   | **12.** Transverse waves in a rod |
|   |   |      | Q |     |   | **13.** Resonant standing waves in a rod |
|   | N |      |   | O   |   | **14.** Standing waves on the surface of a liquid—Part I |
|   |   | $    |   | O   |   | **15.** Standing waves on the surface of a liquid—Part II |
|   | N | ¢    |   |     |   | **16.** Standing waves on the surface of a liquid—Part III |
| ! |   | $    |   |     |   | **17.** Standing waves on a soap bubble |

## List of Demonstrations

! = SURPRISING RESULT
N = NEW DEMONSTRATION
¢ = COST: NEAR ZERO
$ = COST: BETWEEN "NEAR ZERO" AND $2
$$ = COST: BETWEEN $2 AND $12
Q = QUANTITATIVE COMPONENT
O = OVERHEAD PROJECTOR OPTIONAL
R = OVERHEAD PROJECTOR REQUIRED
C = SOME CONSTRUCTION REQUIRED
+ = DURATION: MORE THAN ONE MINUTE
* IN MARGIN = SOME DEGREE OF HAZARD

### K. Doppler Effect and Beats

| ! | N | Cost | Q | O/R | C | + | * | Demonstration |
|---|---|---|---|---|---|---|---|---|
| ! | | $ | | | | | | 1. Whirling a beeper in a circle |
| ! | | $ | | | | | | 2. Shaking a rod vibrating in its longitudinal mode |
| | | $$ | | | | | | 3. Beats using two sound sources |

### L. Electricity

| ! | N | Cost | Q | O/R | C | + | * | Demonstration |
|---|---|---|---|---|---|---|---|---|
| | | $$ | | O | | | | 1. Solar-powered fan |
| ! | | $ | Q | | | | | 2. Lemon battery |
| ! | | $$ | | | | | | 3. Human battery |
| | | $$ | | | | | | 4. Human resistance |
| | | | | O | | + | | 5. Mapping the electric field |
| ! | | ¢ | | O | | | | 6. Charge-propelled aluminum can |
| ! | N | $$ | | O | | | | 7. Light bulbs in series and in parallel |
| | N | | | O | | | | 8. Discharging a capacitor through a light bulb |
| | | | Q | O | | + | | 9. Discharging a capacitor through a voltmeter |
| | | $$ | Q | | | + | | 10. Conversion of electrical to thermal energy |
| ! | | | | O | | + | * | 11. Superconductivity |
| ! | | $ | | | | | * | 12. Twirling a neon lamp on a line cord |
| | N | ¢ | Q | R | | + | | 13. Departures from the inverse-square law |

### M. Magnetism

| ! | N | Cost | Q | O/R | C | + | * | Demonstration |
|---|---|---|---|---|---|---|---|---|
| | N | $ | | O | C | | | 1. Force between two coils |
| | | $ | | O | C | | | 2. Field of a long straight wire |
| | | $ | | O | C | | | 3. Field of a coil |
| | | $ | | O | C | | | 4. Field of a magnet |
| ! | | | | | | | | 5. Deflection of an electron beam |
| ! | | $ | | | | | | 6. Deflection of a light-bulb filament |
| ! | | $ | | O | C | | | 7. The world's simplest motor |

## List of Demonstrations

! = SURPRISING RESULT

  N = NEW DEMONSTRATION

    ¢ = COST: NEAR ZERO

    $ = COST: BETWEEN "NEAR ZERO" AND $2

    $$ = COST: BETWEEN $2 AND $12

      Q = QUANTITATIVE COMPONENT

        O = OVERHEAD PROJECTOR OPTIONAL

        R = OVERHEAD PROJECTOR REQUIRED

          C = SOME CONSTRUCTION REQUIRED

            + = DURATION: MORE THAN ONE MINUTE

              * IN MARGIN = SOME DEGREE OF HAZARD

| ! | N | $ | Q | O/R | C | + | * | | |
|---|---|---|---|---|---|---|---|---|---|
| | | | | | | | | **N.** | **Induced EMF and Lenz's Law** |
| ! | | $$ | | | | | * | **1.** | Bringing a coil with a bulb near an AC-powered coil |
| | | | | | | | | **2.** | Moving a magnet near a coil with a galvanometer |
| ! | | $ | | O | | | | **3.** | Pumping a hanging coil with a magnet |
| ! | | $ | | O | | | | **4.** | Magnet pendulum swinging over a copper sheet |
| ! | N | $$ | Q | | C | + | | **5.** | Dropping a magnet down a pipe |
| ! | | $$ | | | | | | **6.** | Motor driver and slave |
| ! | | $$ | | | | | | **7.** | Hand-powered generator |
| | | | | | | | | | |
| | | | | | | | | **O.** | **Polarization and Electromagnetic Waves** |
| | | $$ | | R | | | | **1.** | Polarization of scattered light |
| | | $ | | | | | | **2.** | Polarization of reflected light |
| ! | | $ | | R | | | | **3.** | Crossed Polaroids and optical activity |
| | | $ | | | | | | **4.** | Wave on a spring through a vertical slit |
| ! | | $ | | O | | | | **5.** | Calcite crystal |
| ! | N | ¢ | | | | | | **6.** | Shielding a radio |
| ! | N | ¢ | | | | | | **7.** | Transverse nature of radio waves |
| | | | | | | | | | |
| | | | | | | | | **P.** | **Geometrical Optics** |
| | N | $$ | | R | | + | | **1.** | Underwater optics |
| | | $$ | Q | | | + | | **2.** | Multiple images between two inclined mirrors |
| ! | | $$ | | | | | | **3.** | Three-corner reflector |
| | | $$ | | O | | | | **4.** | Fresnel lens |
| ! | | $$ | | O | | | | **5.** | "TV rock" (ulexite) |
| ! | N | ¢ | | | | | | **6.** | Total internal reflection at a two-liquid interface |
| ! | N | ¢ | | | | | | **7.** | Fountain of light |
| | | ¢ | Q | R | | | | **8.** | Apparent depth of a submerged object |

## List of Demonstrations

! = SURPRISING RESULT
N = NEW DEMONSTRATION
¢ = COST: NEAR ZERO
$ = COST: BETWEEN "NEAR ZERO" AND $2
$$ = COST: BETWEEN $2 AND $12
Q = QUANTITATIVE COMPONENT
O = OVERHEAD PROJECTOR OPTIONAL
R = OVERHEAD PROJECTOR REQUIRED
C = SOME CONSTRUCTION REQUIRED
+ = DURATION: MORE THAN ONE MINUTE
* IN MARGIN = SOME DEGREE OF HAZARD

| ! | N | Cost | Q | O/R | C | + | Demonstration |
|---|---|------|---|-----|---|---|---------------|
|   | N | ¢ |   | O |   |   | 9. Rainbow formation |
| ! | N | ¢ | Q | O |   |   | 10. Turning the world inside out |

### Q. Interference and Diffraction

**General:**

| ! | N | Cost | Q | O/R | C | + | Demonstration |
|---|---|------|---|-----|---|---|---------------|
| ! |   | ¢ | Q | R |   | + | 1. Superposition of circle transparencies |
| ! |   | ¢ |   | R |   |   | 2. Moire patterns |
|   | N | $$ | Q | R | C | + | 3. Ripple tank |

**Acoustic:**

| ! | N | Cost | Q | O/R | C | + | Demonstration |
|---|---|------|---|-----|---|---|---------------|
| ! | N | $$ |   |   |   |   | 4. Sound through a fan |
| ! | N | $ |   |   | C |   | 5. Diffraction with a rolled-up carpet |
| ! | N | $ |   |   |   |   | 6. Moving a beeper below desk level |
| ! | N | $ | Q |   |   |   | 7. Beeper at the center of a hollow tube |
| ! | N | $$ |   |   | C |   | 8. Acoustic diffraction grating |

**Optical:**

| ! | N | Cost | Q | O/R | C | + | Demonstration |
|---|---|------|---|-----|---|---|---------------|
| ! |   | $ |   |   |   |   | 9. Squinting at an unfrosted bulb |
| * |   | ¢ |   |   |   |   | * 10. Variable-width slit using razor blades |
|   |   | $$ | Q |   |   |   | 11. Diffraction-grating sheets |
| ! |   | ¢ |   |   |   |   | 12. Pin holes in aluminum foil |
|   |   | $ |   |   |   |   | 13. Hologram eyeglasses |
| ! |   | ¢ |   |   |   |   | 14. Interference in thin films |
| ! | N | ¢ | Q |   |   |   | 15. Diffraction rings through a mist |

# Turning the World Inside Out

### AND 174 OTHER SIMPLE PHYSICS DEMONSTRATIONS

# Accelerated Motion and the

# Acceleration of Gravity

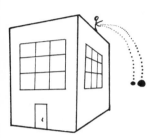

## **A.1.** Dropping balls of different sizes

**Demonstration**

By dropping balls of different sizes from the same height, you can show that they fall together, unless one has an extremely low density or small size. The balls hit the ground simultaneously, even if one is propelled with a sizable horizontal velocity—but only if air resistance is negligible.

**Equipment**

A stick; two small blocks of wood; and a collection of balls of different sizes—for example, a Ping-Pong ball, a tennis ball, and a basketball. You might also include a pair of identical-looking "bounce/no-bounce" balls, available from Toys-R-Us for about $1.00, and two 1-in.-diameter Styrofoam balls obtainable at a crafts store.

**Comment**

As an attention-getter, you might want to start with the "bounce/no-bounce" balls to show that identical-looking balls can have drastically different properties. The observation that balls of different sizes fall together can be conveniently demonstrated by dropping several balls from a common height of about 6 feet, and observing their simultaneous impacts. Be sure to include at least one very light ball—say, a Ping-Pong ball—to see the importance of air resistance. Surprisingly, you will probably not notice any difference between a Ping-Pong ball and a denser ball, although you will notice a sizable difference when you replace the Ping-Pong ball with the small Styrofoam ball. (The circumstances under which air resistance becomes important are explored in demonstration D.6.)

You should also try dropping one object while simultaneously launching another with some initial horizontal velocity. One easy method of doing this uses a stick and two blocks of wood. Place block $A$ near the edge of a desk and block $B$ on top of a stick that projects over the edge of the desk. You can propel block $A$ horizontally by sweeping the stick into it. If block $B$ is originally placed on the part of the stick that projects out beyond the desk, block $B$ falls straight down during the stick's horizontal sweep. Since the blocks begin their descent almost simultaneously, their impacts with the floor should also be simultaneous, since an object's vertical acceleration is independent of its horizontal speed if air resistance is negligible.

You can easily show that the horizontal and vertical motions are *not* independent when air resistance is important by repeating the two-block demonstration using two small Styrofoam balls: the ball propelled at a high horizontal velocity lands after the one that falls straight down, because it experiences a greater retarding force in the vertical direction owing to its higher velocity.

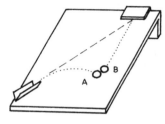

## A.2. "Monkey and hunter" demonstration on an incline

**Demonstration**

You can perform the classic "monkey/hunter" demonstration using two balls, $A$ and $B$, on a slightly inclined surface. If the balls are simultaneously launched, $A$ strikes $B$, because both balls roll the same distance down the incline, regardless of the slope of the incline.

**Equipment**

In the "monkey/hunter" problem, a monkey that lets go of a tree limb on seeing the flash of a hunter's gun is struck by a bullet whose vertical fall exactly matches his own. The equipment you need for this version of the monkey/hunter demonstration includes a 9×12-in. piece of half-inch-thick acrylic sheet; a 3-in.-long piece of transparent plastic corner molding (the "gun barrel") should be glued or screwed onto the lower left corner of the acrylic sheet, and a 2-in. piece of acrylic (the "tree limb") glued or screwed onto the top right corner (see Plate A.2a). You also need two 1-in.-diameter metal balls and two screws to prop up the top end of the sheet. You only need to use an acrylic sheet if you want to give the demonstration to a large group using an overhead projector; otherwise, any opaque sheet of material would suffice. (Acrylic sheets are particularly easy to cut and drill, and they can be purchased from hardware stores or plastics companies.)

Accelerated Motion

**A.2a** "Monkey and
hunter" apparatus

### Construction

It is better to glue the corner molding than to screw it onto
the acrylic sheet;this way one avoids screw heads getting in
the way of the rolling ball. Be sure to glue the corner molding so that it points directly at the center of the top piece of
acrylic where the other ball will be positioned. You may
want to try screws of different lengths to prop up the top of
the sheet, but one inch seems to work fine. A nice touch is
to make a drawing of a hunter's gun and a monkey hanging
from the tree branch on the same scale as the sheet. You can
then make a transparency from this drawing, and tape it on
the underside of the acrylic sheet.

**Comment**

The conventional "monkey/hunter" demonstration requires a large apparatus because the "gun" must be widely separated from the "monkey" in order to give both the projectile and the "monkey" a chance to fall a noticeable distance. It also requires some kind of electromagnet release mechanism to ensure that the monkey lets go at the instant the projectile is launched. By using a slightly inclined plane, which in effect dilutes gravity, you can do the demonstration on a small scale in slow motion. As a result of the slow motion, you can release ball *B* with one hand and launch ball *A* with your other hand almost simultaneously. In principle, the projectile ball should strike the "monkey" for any launch velocity, but don't expect hits 100 percent of the time, particularly for very slow launch velocities, for which simultaneous ball releases become particularly important for success. Incidentally, in the case of the actual monkey/hunter situation, if the monkey is shot at from far enough away for the bullet and monkey to fall an appreciable distance, the bullet would probably miss, because air resistance is quite important for high-velocity bullets, and the vertical and horizontal motions are not independent in this case, as shown in demonstration A.1.

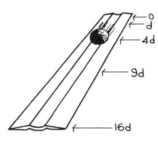

## A.3. Rolling balls down an inclined ruler

**Demonstration**

By rolling a steel ball in the groove of a slightly inclined twelve-inch plastic ruler, you can verify the relationship $s = \frac{1}{2}at^2$ on an overhead projector, and determine $g$, the acceleration due to gravity.

**Equipment**

A ball; a twelve-inch clear plastic ruler that has a groove down the middle; a supply of index cards; and a metronome if one is available (it is not essential). The ball needs to be able to accelerate freely down slopes that are nearly horizontal. A highly polished stainless steel ball of 1-in. diameter has a suitably small coefficient of rolling friction. The twelve-inch ruler should be shortened to 10 inches in order to fit entirely on the glass surface of an overhead projector. Even if you are doing the experiment on your own, and don't need to use an overhead projector, shorten the ruler anyway, so that the numbers given below apply to your situation.

## Accelerated Motion

### Comment

According to the relation $s = \frac{1}{2}at^2$, a ball that rolls a distance $d$ down an incline in 1 second rolls 4 times as far in 2 seconds, 9 times as far in 3 seconds, and 16 times as far in 4 seconds. This rule can be tested by rolling a ball on an inclined ruler, provided the angle of incline is sufficiently small. Obviously, the angle must be small indeed for the rolling ball to remain on a ten-inch ruler for as long as 4 seconds. The angle of incline determines $d$, and we require that $16d$ be less than 10 inches. As shown later, the choice $A = 0.23$ degrees yields a suitable value of $16d = 9$ in. $= 22.86$ cm. Surprisingly, it is not difficult to tilt a plastic ruler so that it has a 0.23-degree angle with the horizontal to a high degree of accuracy (about ±0.02 degrees).

First, you need a transparency to place under the ruler. On the transparency draw five bold horizontal arrows, with the symbols 0, $d$, $4d$, $9d$, and $16d$ next to the arrows, spaced as follows: The zero arrow should be near the top of the transparency and the other four arrows should be located 1.43, 5.72, 12.86, and 22.86 cm below the zero arrow. Place the plastic ruler on top of the transparency on an overhead projector with the ruler next to the five arrows.

In order to tilt the ruler at a precise 0.23-degree angle, you first need to level it accurately. Leveling is done using a supply of 1×3-in. strips cut from index cards, which serve as shims to place under the ruler. Note that if you move either the projector or the ruler after it is leveled it will need to be leveled again. You can level the ruler by placing index-card strips under one end until a ball placed anywhere on the ruler remains at rest. However, since the ruler is somewhat flexible, particularly with the ball's weight on it, you will probably have to place one index-card strip under the middle for every two you place at one end. An even better way to deal with the problem of ruler flexing due to the weight of the ball is to tape the ruler securely to a piece of half-inch-thick acrylic sheet or glass to stiffen it, and tape the arrow transparency to the underside.

Once the ruler is leveled you will want to raise one end by another 1.0 mm, to achieve a 0.23-degree angle. To accomplish this you need to determine the thickness of each index-card strip by measuring the thickness of a large number of cards—say, 50. With my index cards I found each card was 0.2 mm thick, so that to raise the "top" end by 1.0 mm I placed 5 strips under it (in addition to the leveling strips). I didn't need to add 2 or 3 strips under the middle, since I chose to stiffen the ruler by taping it to a piece of acrylic.

After raising the top end by 1.0 mm, the ruler is tilted at an angle of $(0.23 \pm 0.02)$ degrees with the horizontal. (The uncertainty $\pm 0.02$ degrees corresponds to an uncertainty of half the thickness of an index card in the height of the top end of the ruler.) A ball rolling down an incline of angle $A$ has an acceleration $a = \frac{5}{7}g \sin A$, not $g \sin A$, which applies to a frictionless sliding object. Thus, the ball's acceleration is $a = \frac{5}{7}(32)\sin 0.23° = 0.092$ ft/s$^2$. According to the relationship $s = \frac{1}{2}at^2$, the ball will travel a distance of 9 inches at the end of 4 seconds, as stated earlier.

Place the ball at the zero arrow, and release it while you simultaneously begin counting off the seconds ("one one-thousand, two one-thousand, . . ."). You should find that the ball's position at 1, 2, 3, and 4 seconds reasonably coincides with the positions of the four arrows at $d$, $4d$, $9d$, and $16d$, particularly if you look at the rolling ball as you count off the seconds(!). In principle, you can do the same experiment more quantitatively using a metronome adjusted for one second ticks, but getting a precise agreement is difficult given the errors in the ruler's angle and intrinsic flatness. You might try to compensate for these factors by slightly varying the metronome frequency to get more exact agreement.

These observations verify the relation $s = \frac{1}{2}at^2$, and they show that $g$ is close to its nominal value. You could alternatively compute $g$ using a stopwatch by observing the time $t$ required for the ball to reach the $16d$ arrow. You can find $g$ in ft/s$^2$ from the relation $g = 512/t^2$, which for the case $t = 4$ seconds would yield the nominal value $g = 32$ ft/s$^2$. Given an uncertainty of 10 percent in the angle of the ruler, the measurement of $g$ is likely to be accurate to no better than 10 percent, i.e., $\pm 3.2$ ft/s$^2$.

## A.4. Rolling balls on a vibrating plate

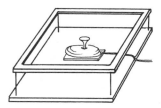

### Demonstration
Balls rolling on a plate vibrating at 60 Hz leave a trail of dots equally spaced in time (1/60 sec apart), providing, at minimal cost, the same type of record as stroboscopic photographs of a puck moving on an air table. For example, by tilting the plate and rolling the ball, parabolic dot trails can be obtained, allowing you to determine $g$.

### Equipment
Steel balls of various masses; a two-speed vibrator/massager (Sears has one for about $20.00); a 9×12-in. wooden picture

frame (or a handmade frame made from corner molding); a 9×12-in. piece of ¼-in.-thick composition board (obtainable at any building materials store); several ⅛-in.-diameter aluminum rods; a 9×12-in. piece of wood; and some pencil carbon paper (more sensitive than regular carbon paper).

**Construction**

The vibrating plate on which you roll the balls rests on a 9×12-in. picture frame, which has a convenient ledge if the back of the frame faces upward. The plate's vibrations are caused by a vibrator/massager placed on a wooden base beneath the plate. The entire apparatus with the plate removed for clarity is shown in the figure. To construct the apparatus, first drill ³⁄₃₂-in. holes in the four corners of the picture frame, and drill correspondingly placed holes in the 9×12-in. wooden base. You should drill the four holes with the picture frame and wooden base clamped together so that the holes are in corresponding positions. The picture frame is supported above the wooden base by four ⅛-in. aluminum rods that will fit snugly in the holes and connect the frame rigidly to the base. You could also use wooden dowels instead of aluminum rods. The plate that you place on the ledge of the picture frame is made from a piece of composition board that must be cut carefully so that there is a ¹⁄₁₆-in. clearance all around. This clearance allows the plate to rest on the ledge without its vibrations being restricted. The ⅛-in. rods should be cut to the length at which, when the vibrator is put on the base, the massager tip just barely makes contact with the plate. (Use a flat tip if the vibrator/massager comes with an assortment of tips.) Tape the vibrator tip securely to the bottom of the plate using duct tape. This important step ensures a uniform vibration amplitude for the plate.

You also need to use four strips of wood to make a four-sided enclosure in which the vibrator can sit snugly directly beneath the center of the plate, because the vibrator would otherwise tend to wander once it is turned on. Place carbon paper face up on the plate, and place plain paper (or a transparency blank if giving the demonstration to a large group) on top of it. The simplest way to obtain the position of the dots as a function of time would be to record the parabolic dot trails directly onto graph paper. If you are using transparencies rather than paper, you will find that the dot trails will be rather faint, so you may wish to prepare in advance some transparencies on which you have darkened every fifth dot. Note that this is just one of a dozen experiments that can be done with balls rolling on a vibrating plate. The entire set of experiments is described in an article in *The Physics Teacher* (*TPT*): R. Ehrlich, "Air Table Experiments Without an Air Table," *TPT 23*, 113–16 (February 1985). In that *TPT* article a magnet coil was used with a steel plate to achieve the vibrations—a method that is not as simple and effective as the method described here.

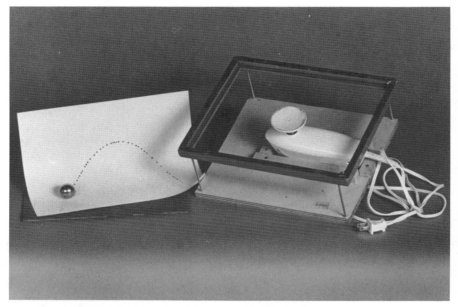

**A.4a** Vibrating-plate apparatus with plate removed for clarity

**Comment**

With the vibrator turned on, roll a steel ball across the paper, and examine the dot trail caused by the vibration-induced pressure fluctuations. The vibrator results in 60 vibrations per second, so the dots are spaced by $\frac{1}{60}$ sec. You will need to experiment with metal balls of different masses, and high/low vibrator settings, in order to see which combination gives the clearest dot trails. You will find that heavier balls give darker trails (important for a transparency), but these trails do not show distinct dots as well as the trails from lighter balls do. Once you have decided how to achieve the most distinct dot trails, roll a ball across the paper and examine its dot trail. If the device is not tilted, the dots should be approximately evenly spaced, indicating that the ball's velocity was constant. Now, tilt one edge of the base up by an inch or two by placing something under it, and measure the angle $A$ of the incline. If you release a ball from rest down the incline (the $y$ direction), the dot spacings will be found to increase with time, since the ball is accelerating. You can also try launching the ball from the bottom of the incline with varying velocities and angles, which with practice will create symmetric parabolic dot trails. A quantitative analysis of either the linear or parabolic dot trails begins with numbering the dots, making an arbitrary choice of the first (time $t = 0$) dot. For each subsequent dot, you can compute the $x$ and $y$ components of its average velocity by dividing its $x$ and $y$ coordinates (relative to the $t = 0$ dot) by the elapsed time, which equals the number of the dot in units of $\frac{1}{60}$ second. For each dot in a trail you can also find the $x$ and $y$ components of the instantaneous velocity by dividing the $x$ and $y$ separations of adjacent dots by $\frac{1}{60}$ second. Make a table showing the $x$ and $y$ components of the average and instantaneous velocities for each numbered dot. At least three ways exist to display these data and find the acceleration in the $y$ direction, $a_y$. (The same methods work to find $a_x$, which should be zero.)

(1) Compute $a_y$ for each pair of adjacent dots by dividing the difference in their instantaneous $v_y$ by $\frac{1}{60}$ second, and plot $a_y$ versus time (dot number).

(2) Plot the instantaneous $v_y$ for each dot versus time, and find the acceleration $a_y$ from the slope of the straight line that most nearly passes through the plotted points.

(3) Plot the *average* $v_y$ for each dot versus time, and find the slope of the straight line that most nearly passes through the plotted points. The acceleration $a_y$ is twice this slope, because the average velocity can be expressed as $v_{0y} + \frac{1}{2}a_y t$.

For ideal (noiseless) data, all three of the preceding methods would give the same results. In practice, however, sizable irregularities in dot spacings cause method (1) to give results that are virtually meaningless, method (2) to give results that have sizable fluctuations, and method (3) to give good results, with smaller fluctuations. Despite the poorer quality of the results from methods (1) and (2), it might nevertheless be worthwhile to try all three methods to see how the same set of data can provide very different results depending on the method of analysis employed. Once you have determined the acceleration $a$ you can use it together with the measured angle of incline, $A$, to determine $g$ using the relation $a = \frac{5}{7}g \sin A$ for the acceleration of a rolling ball on an incline. The uncertainty in your value for $g$ will primarily be determined by the uncertainty in your angle measurement.

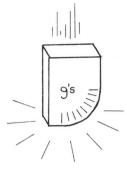

## A.5. Dropping an accelerometer from various heights

**Demonstration**

Dropping an accelerometer onto different surfaces and from different heights shows qualitatively the dependence of deceleration on the initial speed and the stopping distance.

**Equipment**

An "impact-stress meter" made by RunTronics, Inc. (see address in Appendix 2), sold for about $20.00. This device emits a high-pitched beep when it is held vertically and decelerated at a rate greater than some value—ranging between 2.3 and 5.0 $g$'s—that can be set on the dial. (The device helps runners use biofeedback to achieve a soft landing each step; runners adjust their stride until the device stops beeping with each impact.) An alternative device, in case you can't obtain the RunTronics product, would be an "impact indicator" made by Index Packaging, Inc., which costs about $4.00. This device sets a colored indicator when the device is subject to some particular acceleration such as 5 $g$'s. It can be affixed to packages to see whether they are handled too roughly while in transit.

**Comment**

Set the acceleration threshold value at 2.3 $g$'s and drop the accelerometer from one hand, catching it in the other after it falls a short distance. Because this is a relatively "hard" landing, you will not be able to drop the device very far before it beeps. Dialing the highest value (5.0 $g$'s) will allow you to drop it from a somewhat greater height. Now try dropping the device and catching it while bringing it to rest gradually, i.e., "breaking" its fall or increasing its stopping distance. You will obviously find that the device can now be dropped from greater distances before it beeps at 2.3 $g$'s and 5.0 $g$'s. Unfortunately, a quantitative comparison is not possible when one cannot control the stopping distance.

# Section B

## Gravity and Curved Space-Time

### B.1. Rolling balls on a stretched membrane

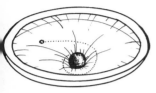

**Demonstration**
The curved space-time interpretation of gravity in relativity theory can be simulated by rolling balls on a stretched transparent membrane placed on an overhead projector.

**Equipment**
A 10-in.-diameter embroidery hoop (preferably the plastic type, obtainable at a crafts or fabric store); stretchable plastic sandwich wrap (Handiwrap is better than Saran Wrap); a one-inch-diameter steel ball; and some BB's (obtainable at most hardware or gun stores). You might want to use the ¼-in.-diameter BB's used in slingshots, which are easier to handle than the smaller ones used in BB guns.

**Construction**
Put the stretched plastic wrap on the embroidery hoop, being sure to follow any instructions that come with the hoop. Try to stretch the wrap so that you get all the wrinkles out.

**Comment**
According to general relativity, gravity is not due to a force, but rather is due to the curvature of space-time caused by the presence of matter, and that curvature or distortion affects how other matter moves. To simulate how this works try the following three demonstrations:

(1) Roll a BB across the surface of the stretched membrane. (Be sure the lip of the embroidery hoop faces up; otherwise you will be chasing BB's all over the floor.) With the hoop supported at its edges the rolling BB causes a slight local distortion of the membrane, but it travels in a straight line because no *other* mass has caused an overall distortion of the membrane.

(2) Now place a one-inch steel ball at the center of the membrane, causing the membrane to be appreciably distorted. If you roll BB's from the edge of the membrane they will either "go into orbit" about the center ball if their speed is low enough, or else be deflected by the curvature of the

membrane and strike the hoop. Using different launch speeds and directions, you can achieve elliptical orbits with various eccentricities. You will find that the orbits precess, i.e., show a rotation of the major axis of the ellipse on successive orbits. The precession is due to the non-inverse-square nature of the force acting on the BB's. A planet subject to the inverse-square gravitational force of the sun exhibits much less precession, owing primarily to the perturbations from the other planets. In the case of the planet Mercury, a small unexplained precession was a key confirmation of Einstein's general theory of relativity. Unlike the case of planetary orbits, in our simulation the orbits decay very rapidly owing to the inelasticity of the membrane. (Membranes made out of a rubber sheet are more elastic than plastic wrap, but if you want to do the demonstration using an overhead projector you would need to use a piece of thin transparent latex.)

(3) The formation of stars, planets, or galaxies due to the mutual gravitational attraction of many separate particles for one another can be simulated using some number of BB's (say 10) placed at random points around the membrane. Now, gently shake the hoop trying to avoid letting the BB's strike the edges of the hoop with too much force. Because of their random motion and the small local membrane distortions at each BB, soon the BB's all coalesce and move about the membrane as a single mass, provided the hoop is not shaken too vigorously.

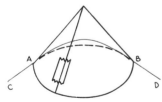

## B.2. Simulation of the gravitational deflection of light

### Demonstration
The bending of light rays by gravity can be explained as an effect of curved space-time, and simulated using overhead transparencies. The simulation is a model for the deflection of light rays by the sun, which always occurs but is observable only during a solar eclipse.

### Equipment
An overhead transparency, if you want to give the demonstration on an overhead projector. (Note that most photoduplicating machines can be used to make transparencies from drawings.) Ordinary paper can be used if you have no need for an overhead projector.

**Gravity and
Curved Space-Time**

**Construction**

In this demonstration a cone rather than a stretched membrane is used to simulate a region of curved space-time. The massive body responsible for the space-time distortion is assumed to lie at the apex of the cone. The continuous path *CABD*, part of which lies on the cone, represents the path of a light ray. To construct the cone first make a transparency from a photocopy of Plate B.2a. Cut the circle out of the top part of the transparency and cut the 30-degree wedge out too. Bring the edges of the wedge-shaped gap together to make the plane figure into a cone with the center of the circle as the cone's apex. Note that the original straight line segment *AB* is no longer a straight line once the plane figure is made into a cone. Make the cone into a rigid structure by taping its seam together with transparent tape. Now place the cone onto the bottom part of the transparency, so that the line segment *AB* joins smoothly onto the lines labeled *C* and *D*, and place both on an overhead projector. You can verify that the lines join smoothly by pressing the edge of the cone flat at points *A* and *B*.

**Comment**

In general relativity, light rays travel only in straight lines in flat space-time. In a space-time that is distorted by the presence of matter, light rays travel along the shortest possible path, known as a geodesic. Obviously, the departure from a straight-line path is greater the more massive the body causing the space-time distortion. We can simulate the deflection of a light ray by a massive body by artificially assuming that the distortion is restricted to an area of specific radius around the massive body—the radius of the base of the cone. The straight line path *AB* on the original flat figure becomes a geodesic once the plane figure is made into a cone. Beyond the base of the cone, space-time is assumed flat and the light ray follows straight-line paths *CA* and *BD* (geodesics in flat space-time). Thus, the entire path of the light ray from *C* to *D* consists of a three-piece geodesic. When the two transparencies are placed on an overhead projector, you should be sure to press the cone down at points *A* and *B* to show that the curve *AB* is tangent to straight lines *CA* and *BD*.

One limitation of this demonstration is that the distortion of space-time around a massive body is not restricted to a region of fixed radius but extends outward indefinitely, making the stretched membrane simulation more realistic. In addition, both this demonstration and the stretched membrane demonstration suffer from two important limitations:

# Section B

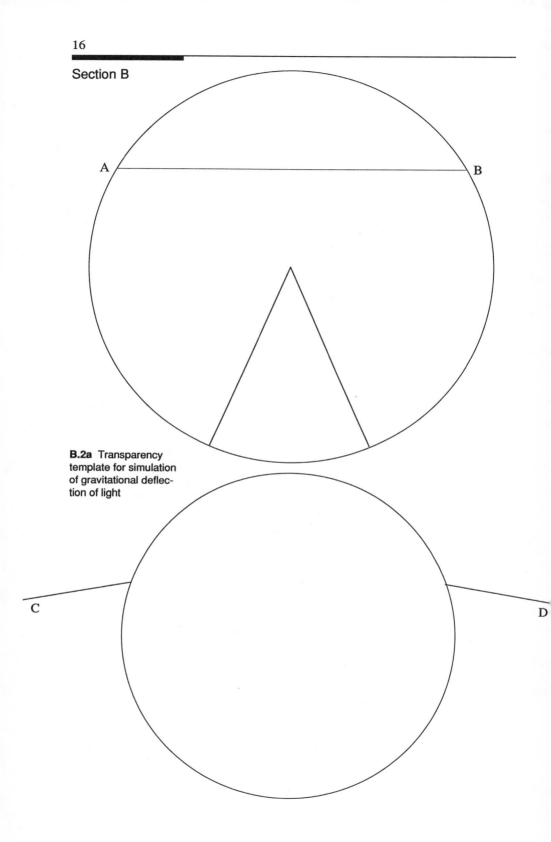

**B.2a** Transparency template for simulation of gravitational deflection of light

**Gravity and
Curved Space-Time**

First, they both treat space as though it were two-dimensional, since we three-dimensional creatures are incapable of visualizing a distorted three-dimensional space without being able to view it from a fourth dimension. Second, and more important, both this and the preceding demonstration only illustrate the distortion of space, and fail to include the distortion of the time dimension. The time distortion accounts for fully half the deflection of a light ray by a massive object and is the primary factor for objects having speeds less than the speed of light. The next demonstration explicitly includes the effect of a time distortion in order to show how general relativity accounts for the acceleration of objects by gravity. Both demonstrations B.2 and B.3 were described in the book *Relativity Visualized* by Lewis Carroll Epstein (San Francisco: Insight Press, 1985).

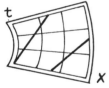

(a)

(b)

## B.3. Acceleration as an effect of curved space-time

**Demonstration**
Relativity attributes the gravitational acceleration of objects primarily to the curvature of the *time* dimension, an effect that can be simulated using overhead transparencies.

**Equipment**
Two transparencies made from photocopies of Plates B.3a and B.3b, which are reproduced at full size on the following pages.

(c)

(d)

**Comment**
Figure (a), left, and Plate B.3a represent a space-time grid corresponding to a flat space-time. Note that the time coordinate is on the vertical axis and the space coordinate on the horizontal axis. The space-time is flat because no nearby matter is present to distort it. The dark slanted lines in Figure (a) represent the path of an object in the flat space-time. Their constant slope represents motion at constant velocity—the velocity being proportional to the *inverse* of the slope of the lines. Unlike our actual space-time, here we pretend space is cyclical: when this "world line" moves off the top of the grid, we show it coming in from the bottom at exactly the space coordinate it went off. When the transparency from Plate B.3a is rolled up as a cylinder [Figure (c)], the slanted line segments all join, and the world line of the moving particle becomes a helix of constant pitch. But just

Section B

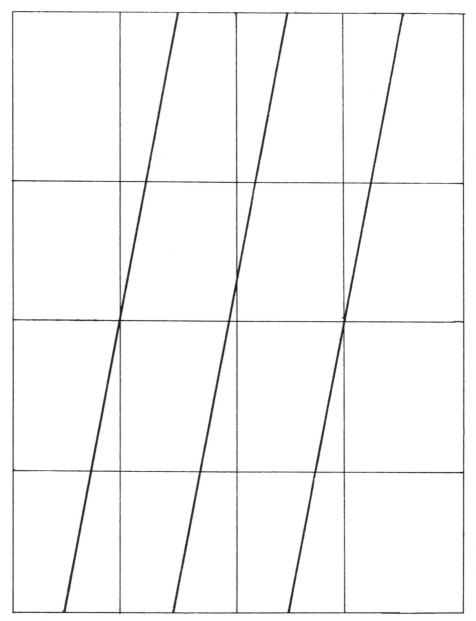

**B.3a** Transparency template for worldline in flat space-time

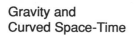

# Gravity and
# Curved Space-Time

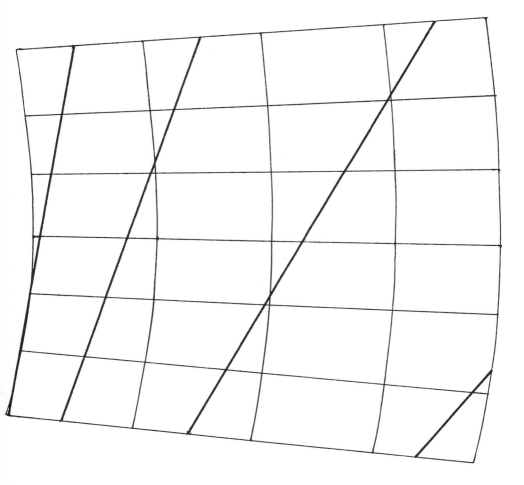

**B.3b** Transparency template for worldline in curved space-time

because we have "rolled up" space-time we have not introduced any curvature or distortion to the grid, which still corresponds to a flat space-time. We can now show how a distortion of the time dimension is responsible for the acceleration of an object due to gravity. Figure (b) (Plate B.3b) represents a grid corresponding to a region of curved space-time. (Actually, only the time axes are curved here). This distortion would be caused by a massive object to the *right* of the figure. According to general relativity, a free particle would travel in a straight line (the slanted line in the figure) even though the space-time grid is distorted. Note that these slanted lines make a larger and larger angle with respect to the time direction the more they are extended, which means that the object's speed is increasing with time. Again the figure can be rolled up, but this time it makes a truncated cone [Figure (d)]. The world line of the particle now becomes a helix of continually increasing pitch, corresponding to an accelerated motion to the right (toward the massive body responsible for the distortion of the space-time grid.) The acceleration was "caused" by a space-time distortion, not by any force—at least, that's the way the situation would be described in general relativity. The perspective of general relativity—free objects travel in straight lines in a curved space-time—is just the opposite of the way classical physics views the situation. In classical physics a free accelerating object, which obeys the equation $x = \frac{1}{2}at^2$, would be represented by the graph of a (curved) parabola on a (straight-line) coordinate grid. Although general relativity is far more difficult mathematically than classical Newtonian physics, it is conceptually simpler, in that there is no need for the concept of a force, and it gives results that are more accurate. Nevertheless, Newton's laws continue to play an extremely important role in physics, and except for situations involving very strong gravitational fields or very high speeds, they accurately describe phenomena.

# Newton's Laws

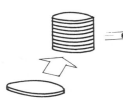

## C.1. Shooting a penny out from under a stack of pennies

**Demonstration**

A penny shot by hand toward a stack of pennies will knock the bottom penny out without disturbing the rest of the stack, but only if the penny's speed is high—an illustration of Newton's first and second laws.

**Equipment**

Five to ten pennies on a smooth surface, such as an overhead projector.

**Comment**

The reason it is necessary to shoot the penny at high speed is that the effect of a given friction force $F$ between the bottom penny and the one above depends on the "impulse" of that force, i.e., the product of the force and the contact time, $F \cdot \Delta t$. The impulse is smallest when the bottom penny is knocked out quickly, causing the least disturbance to the rest of the stack. The minimum speed the penny must have depends on the number of pennies in the stack, because the higher the stack, the higher the friction force, and therefore the smaller the contact time $\Delta t$ to produce the same impulse $F \cdot \Delta t$.

This demonstration does not work every shot, because a penny shot by hand sometimes doesn't hit only the bottom penny in the stack. This problem can easily be avoided, however, if the shot coin has a smaller width than the coins in the stack—for example, shoot a penny at a stack of nickels, or a dime at a stack of pennies.

## C.2 Catching a row of pennies on your arm

**Demonstration**

Newton's first and second laws can be demonstrated by placing a row of 10 to 20 pennies on a plastic ruler laid on your forearm. In one quick motion you can swing your arm forward and catch all the pennies in midair!

**Equipment**
A plastic ruler and 10 to 20 pennies.

**Comment**
The plastic ruler is not essential, but without it you may find it tricky to line up 10 to 20 pennies along your arm and not have some of them slide off. You obviously need to start with your forearm horizontal and swing it rapidly so that your open hand moves roughly horizontally and very quickly. Your hand exerts a force on the ruler, accelerating it forward from under the pennies. The force on the pennies (gravity) is much smaller than that on the ruler, and their smaller acceleration results in a row of pennies in midair, since they have so little time to fall. Your rapidly moving hand can easily scoop up the nearly horizontal row of pennies. (If you execute the arm movement in 0.1 seconds, the farthest penny falls only about 2 inches.) Actually, the slight downward movement of your hand as it moves forward will compensate for this fall of the more distant pennies. Clearly, this demonstration takes a bit of practice, and you may want to start with only a few pennies. It would also be advisable not to have anyone in the direction of your arm swing.

## C.3. Chain with a suspended weight

**Demonstration**
The value of $A$, the angle that each half of a chain supporting a weight makes with the horizontal, depends on how hard the ends of the chain are pulled—an illustration of the first condition for equilibrium (vanishing vector sum of the applied forces).

**Equipment**
A chain; a 10-pound weight from a weight-lifting set; and two demonstration scales. If you don't have a weight-lifting set, any weight tied to the chain will do.

**Comment**
A qualitative version of this demonstration can be done without the two scales if you observe how the angle of the chain with the horizontal decreases as its ends are pulled with increasing force. If you pull the two ends of the chain

as hard as you can, you cannot make the chain appear horizontal if a 10-pound weight hangs in the middle. (A 100-pound pull, which would not be easy to achieve, would still leave a 6° angle in each half.)

You can make the demonstration more quantitative if you pull on scales attached to the chain, and observe how the angle $A$ varies as you vary the pull on the scales. Actually, it is preferable to use a rope or string instead of a chain for the quantitative version so that the weight can slide freely. You should find that for a weight $W$ hung at the middle of the rope, the scale readings $S$ and the angle $A$ are related by $S = W/(2 \sin A)$, as required by the first condition for equilibrium. The concept that no finite force can make the rope exactly horizontal can easily be explained by seeing how the angle varies as the applied force increases. (Of course, the rope can be made momentarily horizontal by suddenly applying the force, causing the weight to shoot upward, because in that case an acceleration is accompanied by unbalanced forces, according to Newton's second law.)

A variation on the demonstration uses unequal angles in the two ends of the rope. For example, using a one-kilogram mass ($W = 9.8$ N) hung from the middle of a rope whose ends make angles with the horizontal of 37° and 53°, respectively, you should find scale readings of 7.8 N and 5.9 N to make the vector sum of the forces on the weight vanish. A convenient way to ensure that the ends of the rope make the correct angles with the horizontal is to lay out the angles on a piece of paper in advance, and make the two ends of the rope parallel to the lines on the paper taped to the blackboard directly behind the rope.

## C.4. Force table

**Demonstration**
A novel type of force table, suitable for use on an overhead projector, shows the relationship between three forces in equilibrium

**Equipment**
An 8-in.-diameter circular disk cut from a piece of half-inch-thick acrylic plastic sheet; 12 large nuts to use as weights; and a 4-in.-long ⅛-in.-diameter machine screw with four nuts and washers. A transparent plastic plate or a plastic salad cover would be suitable alternatives to the acrylic disk.

## Construction

If you do not have a ready-made plastic disk in the form of a plate or salad cover, saw an 8-in.-diameter disk from an acrylic sheet, making it as round as you can. Locate the disk's exact center by balancing it on a nail, and drill a ⅛-in.-diameter hole there. Make transparencies from photocopies of Plates C.4b and C.4c, and cut out each of the three circle shapes. Put the 4-in.-long screw through the hole in the acrylic disk, and also through the centers of the three transparencies placed on top of the disk. Four nuts and two washers should be arranged on the machine screw as shown in part (e) of Plate C.4c. The two nuts at the bottom of the screw should be locked in place. If you drilled the hole through the disk's exact center, you should find that the disk easily balances in a horizontal position on the bottom nuts. (Without the bottom nuts, the screw's smaller surface area would not allow you to balance the disk.) Turn the screw through the locked nuts so that a *very* slight amount projects through the bottom nut. The amount of projection is correct when you can still (barely) balance the disk, but not as easily as before. Now take three lengths of string, and at one end tie 3, 4, and 5 large nuts to be used as weights, and make loops at the other ends. The length of each string should be such that when the loop is put over the screw in the disk, the nuts just hang over the edge of the disk (see Plate C.4a).

**C.4a** Force-table apparatus

# Newton's Laws

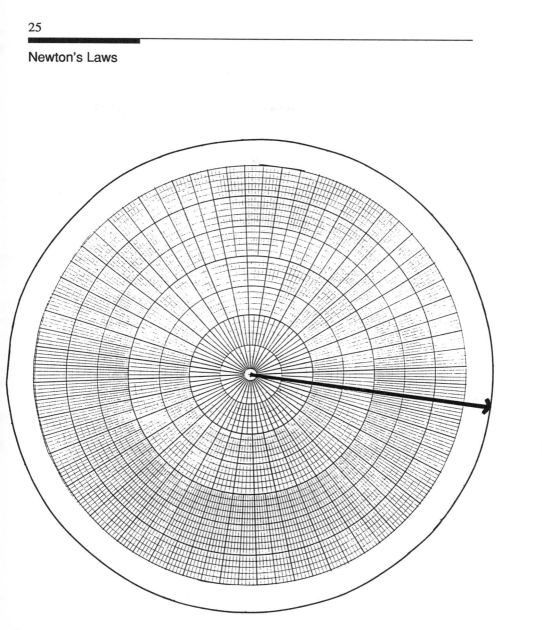

**C.4b** Transparency template for "5-Nut" vector

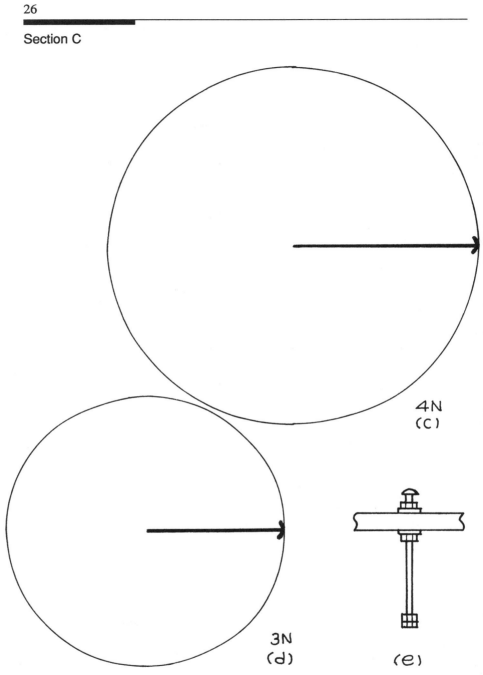

4N
(C)

3N
(d)

(e)

**C.4c** Transparency template for "3-Nut" and "4-Nut" vectors

**Comment**

The arrows on the three transparencies represent force vectors of magnitude 3, 4, and 5 "Nuts," where a force of one Nut equals the weight of a nut. Rotate the three transparencies so that the three vectors are in an equilibrium configuration, i.e., the 3-Nut and 4-Nut vectors make a 90° angle, the 3-Nut and 5-Nut vectors make a 127° angle, and the 4-Nut and 5-Nut vectors make a 143° angle with each other. Now rotate the hanging weights so that the three strings lie along the corresponding vectors. You should find that the disk balances only when the strings lie along or near the three vector directions. The alignment may not be exact if the disk is not exactly circular, if you didn't drill the hole in the exact center, or if the weight of the strings is not negligible. Surprisingly, the large friction force between each string and the edge of the disk is not a source of error, because unlike a conventional rigid force table, the disk tips over whether the unbalanced force is due to a hanging weight or to a friction force acting at the disk's rim. If you want to try balancing weights other than 3, 4, and 5 Nuts, you need to make additional transparencies with the appropriate force vectors.

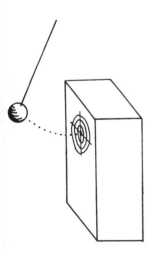

## C.5. Swinging two balls into a block

**Demonstration**

According to the law of conservation of momentum, when two balls having equal momenta collide successively with a block, the one that recoils with greater momentum imparts more momentum to the block, demonstrated by the fact that only this ball knocks the block over.

**Equipment**

A 4×4-in. block of wood; a large Super Ball, made by Wham-O; a lead (or steel) ball embedded in a piece of clay, having the same mass as the Super Ball; some string; and some wire. The proper height of the 4×4 block depends on the mass of the balls: It should be tall enough so that the Super Ball is capable of knocking it over when swung through a 90° arc at the end of a 0.75-meter-long string.

**Construction**

Using the wire, make a holder that tightly encloses the Super Ball so that it can be attached to the string and swung without coming loose. (An alternative would be to glue the string directly onto the Super Ball.) Connect one end of a 1.5-meter-long string to the Super Ball, and connect the other end to a ball of equal mass made from a lead or steel ball embedded in a spherical lump of clay. Make a loop in the center of the string to put your finger in.

**Comment**

You should demonstrate that the two balls have equal masses by hanging them from the ends of a meter stick while balancing the stick on your finger at the 50-cm mark. Now hold the string loop with your finger and release the Super Ball from a 90° angle, swinging it into the block (see illustration). By trial and error, find out at what height on the block the Super Ball must strike in order to barely knock the block over, and make a "bull's-eye" there. When you swing the clay ball (starting with the same 90° angle) toward the bull's-eye, it will give the block only half the momentum of the Super Ball and will not knock it over. The factor-of-two difference in momentum for balls with equal mass and velocity arises because the Super Ball changes its momentum from $+mv$ to $-mv$ (a change of $-2mv$), while the clay ball changes its momentum from $+mv$ to $0$ (a change of $-mv$). If the clay ball tends to stick to the block you may want to wrap it with tape or aluminum foil, so that no one will think the reason the clay ball doesn't knock the block over is that it sticks to it. Another common misconception is thinking that the clay ball doesn't knock the block over because the force on the block is reduced owing to the shock-absorbing quality of clay. It is true that the force $F$ exerted by the clay ball is less, but the collision time $dt$ is greater, and it is the product $F \cdot dt$ (the impulse) that determines the momentum given to the block. As noted previously, the Super Ball gives twice as much momentum to the block as the clay ball, so the impulse is also twice as great for the Super Ball.

An alternative way to conduct this demonstration, which works just as well, is to use a toy Ping-Pong-ball gun fired at a point near the top of a wooden beam balanced on one end. If you tape a sponge near the top of the beam, you can observe what happens when you shoot the ball into the sponge, and then see what happens when you shoot the ball against the beam itself, with the sponge flipped out of the way. Obviously, you need to use a wooden beam of such a length that the Ping-Pong ball shot against the beam (with the sponge flipped out of the way) barely knocks it over.

# C.6. Accelerating a scale with a suspended weight

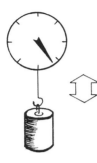

### Demonstration
When a weight suspended from a large demonstration scale is moved up and down, the scale readings permit a quantitative test of Newton's second law.

### Equipment
A large demonstration scale and a weight. Choose a weight appropriate to the range of the scale. You may want to tie the weight on so it doesn't fly off.

### Comment
Consider the following three upward motions: (a) upward at constant speed, (b) upward while accelerating, (c) upward while decelerating. For the three upward motions, you would find that the scale reading equals the weight in case (a), exceeds it in (b), and is less than the weight in (c). These observations can be explained by Newton's second law, according to which the scale reading should equal $mg + ma$. The scale reading, therefore, equals the weight when $a$ is zero, exceeds it when $a$ is positive, and is less than the weight when $a$ is negative. These results apply equally to downward motions, but remember that a downward *deceleration* corresponds to an upward (positive) acceleration.

One complication you face in trying to verify the above predictions is that it is impossible to achieve a sizable constant acceleration for an appreciable length of time when moving the scale and weight by hand. It is easy, however, to move the scale with a varying acceleration that is roughly constant during two halves of the motion. For example, suppose you move the scale and weight upward from point $A$ to point $B$ located one meter higher, and have the scale at rest at the start and finish of its motion. The scale reading should exceed the weight during the first part of the motion, when $a$ is positive, and it is less than the weight during the second part of the motion, when $a$ is negative. If you make the entire motion last one second, and try to make the acceleration and deceleration last equal times, the average acceleration is $+4$m/s$^2$ (0.41 $g$'s) during the first half-second, and $-4$m/s$^2$ during the second half-second. Thus, the scale reading should exceed the weight by 41 percent during the acceleration, and be 41 percent less than the weight during the deceleration. If you try to verify this prediction quantitatively, remember that the acceleration achieved depends on the inverse square of the duration of the motion according to $a = 2s/t^2$, and $t$ cannot be precisely kept to 1 second.

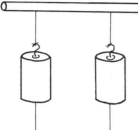

## C.7. Pulling a thread attached to a hanging weight

**Demonstration**
When the bottom string on each weight is pulled downward, Newton's second law requires that the top string break first if the pull is slow, and the bottom string break first if it is fast.

**Equipment**
Two weights; some thread (rather than string); and a wooden or metal bar. Choose weights such that the thread supporting them has a breaking point between 2 and 4 times the weight.

**Comment**
Hold the bar with one hand and pull one of the bottom strings very slowly. This will cause the top string to break. Then pull the bottom string on the other weight rapidly. This time the bottom string will break instead of the top string. The explanation follows from Newton's second law, according to which the tension in the top string, $T_1$, and that in the bottom string, $T_2$, are related by $-T_1 + T_2 + mg = ma$. Thus as long as the mass is momentarily accelerated downward with an acceleration $a$ less than $g$, then $T_1$ exceeds $T_2$, and the top string breaks. If $a$ exceeds $g$, however, then $T_2$ exceeds $T_1$, and the bottom string breaks.

## C.8 "Vampire killer"

**Demonstration**
The absence of any pain when a heavy, pointed stake is put against your chest and struck with a hammer is explained by Newton's second law.

**Equipment**
A stake made from a piece of steel stock weighing around ten pounds; and an ordinary hammer (not a sledgehammer!). The point of the stake should be rounded with a file. An

alternative demonstration uses a lead or steel brick or a 20-pound weight from a weight-lifting set.

**Comment**
Even though vampires are usually killed while they are lying down, you should be standing and not lying down when the stake is struck. Be sure the person swinging the hammer doesn't miss the stake, and don't let the person swing with full force until you know how much it will hurt. If you (a) wear a sweater, (b) use a heavy stake, and (c) don't place the stake at a bony part of your chest or at your solar plexus, then it hardly hurts at all. You feel little pain because, by Newton's second law, the difference between the forces on each side of the stake equals the product of the mass and acceleration of the stake: $T_1 - T_2 = ma$. $T_2$ (the force you feel) is much less than $T_1$ (the force the hammer exerts) because the mass $m$ is large, and so is the product $ma$. On the other hand, if the stake were light, such as the wooden ones actually used to kill vampires(!), the product of mass times acceleration would be much less, and the forces on both sides of the stake would be more nearly equal. Also, if you were lying down, and the stake couldn't move very far, or if your chest had no "give," the forces $T_1$ and $T_2$ would again be more nearly equal—in this case, because the stake's acceleration is reduced. For those who are squeamish about hitting a stake placed against their chest, another form of this demonstration involves having someone strike a lead or steel brick resting on your hand with a hammer. This, of course, allows you to swing the hammer yourself.

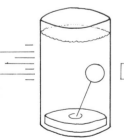

## C.9. Ping-Pong-ball buoy anchored to a weight in a jar

**Demonstration**
When a water-filled jar containing a Ping-Pong ball anchored to a weight is suddenly moved, the floating Ping-Pong ball behaves as though it has a negative mass, and its angle of swing allows the device to function as an accelerometer.

**Equipment**
A wide-mouth jar; a Ping-Pong ball; and a metal weight that preferably fills most of the bottom of the jar (so that it won't rattle around when you accelerate the jar).

## Construction

Glue one end of the string to the Ping-Pong ball and tie the other end to the center of the weight. The longer the string, the smaller the maximum acceleration to which the device responds before the Ping-Pong ball strikes the side of the jar. If you want to measure accelerations of up to 1 $g$, the Ping-Pong ball must swing to an angle of 45° without hitting the side of the jar. You might want to use several jars with different string lengths. If you can't find a nice disk-shaped metal weight that nearly fills the bottom of the jar, you can either epoxy the string onto the bottom or tie it to a bunch of long nails arranged in a cross that will occupy the entire bottom of the jar and not rattle around.

## Comment

The Ping-Pong ball behaves as though it has a negative mass because when the jar is accelerated to the right with acceleration $a$, the water is thrown backwards. The water pressure on the left side of the ball is then higher than that on the right, and the ball swings to the right. Another way to describe the situation is to note that in a reference frame accelerating with the jar, the resultant combination of a horizontal inertial force, $-ma$, and a vertical gravitational force, $mg$, produces a net force on any object, and this net force makes an angle with the vertical given by $\tan^{-1}(a/g)$. As stated earlier, the string's angle with the vertical will be 45° for the special case $a = g$, according to this equation.

Observe the angle the string makes for different types of motion: constant velocity, large and small acceleration, circular motion, and simple harmonic motion. Note that the Ping-Pong ball leans *toward* the center during circular motion. Demonstration G.2 shows how you can make a quantitative measurement with this device when it is made to undergo simple harmonic motion.

## C.10. Throwing eggs at a sheet

### Demonstration

Raw eggs can be thrown with full force at a sheet without breaking, illustrating Newton's second law.

### Equipment

Some raw eggs; a bed sheet; a bowl; and a Ping-Pong ball.

**Comment**

Throwing raw eggs full-force at a sheet illustrates that (a) eggs are sturdier than you might think, and (b) according to Newton's second law, the force on them is not too large as long as they are not brought to rest too abruptly. Be sure to have two people hold the sheet with the bottom part folded upward to catch the eggs after they strike the sheet. You should be able to throw the eggs as hard as you want without breaking them. After throwing your last egg, demonstrate that they are indeed raw by breaking one in the bowl. As a joke, you might then take a concealed Ping-Pong ball (which momentarily resembles an egg) and throw it at the group watching the demonstration.

## C.11. Water rocket, rocket balloon, and balloon-powered helicopter

**Demonstration**

Toy water rockets, balloons, and balloon-powered helicopters are accelerated by reaction forces, thereby demonstrating Newton's second and third laws

**Equipment**

A toy water rocket and "rocket balloons," available from Toys-R-Us; and a balloon-powered helicopter ("whistling space copter"), available from Flight Promotions, Inc. in Toms River, New Jersey.

**Comment**

The water rocket is perhaps the best of the three toys for demonstrating Newton's third law. You may find it surprising that a rocket can be made that uses only water and compressed air as the propellant. The reaction force exerted by the propellant back on the rocket equals the product of the exhaust velocity and the rate at which mass is ejected. You can demonstrate the importance of both the exhaust velocity and the mass ejected by (a) pumping the rocket to various air pressures (by varying the number of times you pump it) and (b) first using only air and then using a mixture of air plus water. The water-air mixture produces a higher rocket thrust than that produced by air alone, because of the greater mass ejected. Although the rocket is meant to be fired straight up, this would not be a good idea indoors, so be sure you aim it so that it has the longest flight path and won't hit anyone.

The less messy rocket balloons illustrate Newton's law of action and reaction just as well as the water rocket, but they can't be used to show the effect of varying the mass ejected. A water-filled balloon cannot take off like a rocket, because even though the ejected water produces more thrust than the air from an air-filled balloon, the greater thrust is more than offset by the much greater weight of the water-filled balloon.

The balloon-powered helicopter is also interesting because of its seeming defiance of Newton's third law: the balloon stem faces upward, and so it might seem that the copter should be driven downward. The airflow, however, is actually directed through tubes in the copter blades and exits downward, which simultaneously generates the upward lift and causes the blades to spin.

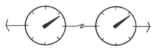

## C.12. Pulling two scales connected together

### Demonstration
Pulling on two connected scales shows that they always read the same value if connected directly, but not if they are connected by a heavy chain, illustrating Newton's second and third laws.

### Equipment
Two large demonstration scales, and a piece of heavy chain.

### Comment
Be sure that the scales have been calibrated to read zero with no force applied, and that both are held horizontal. Even when all the pull "comes from one side," and one scale is connected to a fixed object, the scales still give identical readings. However, this is not the case if you connect the scales with a very heavy chain resting on a table. In this case, the scale readings are momentarily unequal if one scale is suddenly tugged. The acceleration of the chain means that unbalanced forces act at its two ends, according to Newton's second law.

## C.13. Recoil force in a bent straw

### Demonstration
According to Newton's third law, when air is blown through a bent straw, the straw recoils just like a rotating lawn sprin-

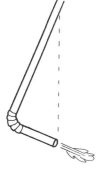

kler. Surprisingly, sucking in air results in no recoil in the opposite direction.

**Equipment**
A flexible soda straw (taped to make the bend a permanent 90° angle).

**Comment**
When you blow hard into the long part of the straw suspended vertically, the free bottom end recoils as indicated in the illustration. In order to convert a downward airflow into a sideways airflow, the straw must exert a force on the air that has a component to the right. The reaction to this force is responsible for the straw's recoil to the left.

It might be expected that if you were to suck in air rapidly instead of blowing it out, the straw would deflect in the opposite direction. Strangely, no deflection is seen, as pointed out by Leonardo Hsu in the April 1988 issue of the *American Journal of Physics* (pp. 307–308). Following an argument by Richard Feynman, Hsu points out that the sucking action results in two horizontal forces on the elbow of the straw, which cancel each other: (1) the force to the right due to the lower pressure inside the straw than outside, arising from Bernoulli's principle, and (2) the impact force to the left when the inrushing air hits the elbow of the straw in making its right angle turn.

It should be noted that other authors have claimed different results for their variations on the inverse lawn sprinkler demonstration. In fact, plausible arguments have been advanced for all three possibilities: no motion, motion same as the normal sprinkler, and motion opposite to the normal sprinkler! See, for example, the article by Richard E. Berg and Michael R. Collier, who report on a demonstration in which a deflection opposite to that of the normal sprinkler is observed. Their article in the *American Journal of Physics* **57**, 654 (July 1989) also gives a summary of the history of this intriguing problem.

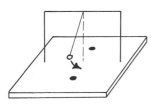

## C.14. Chaotic motion of a pendulum

**Demonstration**
A pendulum swinging above two magnets illustrates that Newton's laws do not always lead to predictable orderly motion—a surprising result from recent studies in the new science of chaos.

## Equipment

Two small disk-shaped magnets, a pendulum with a steel bob such as a paper clip, and an acrylic sheet (if you want to give the demonstration on an overhead projector).

## Construction

Glue or tape the magnets to the top of the acrylic sheet, spacing them by about three inches, and hang the pendulum bob from a horizontal bar connected to two vertical supports attached to the sheet. Leave a small clearance between the swinging bob and the two magnets.

## Comment

Release the pendulum bob from various initial positions and observe its motion. When the bob is released from points in the immediate vicinity of one magnet it will eventually come to rest near that magnet. When the bob is released from more equidistant points its motion will be more complex, and where it eventually settles down seems to be less predictable. By applying Newton's laws it is, however, possible in principle to determine where the ball should come to rest for any given initial position. If the initial position of the bob is in the shaded (unshaded) region of the plane shown in Plate C.14a, the bob will come to rest next to the right (left) magnet.

A magnified view of the boundary between the shaded and unshaded regions, shown in Plate C.14b, reveals that the boundary curve is not smooth, but instead has a "fractal" geometry. A fractal curve, by definition, has a complex structure when viewed under any magnification. A consequence of the fractal character of the boundary is that while one release point for the bob may lead to eventual capture by the right magnet, a second point an infinitesimal distance away may lead to capture by either the right or the left magnet, depending on the exact size and direction of the infinitesimal displacement. Thus, a prediction of where the bob will come to rest requires knowing its initial position with a physically impossible precision, for initial positions near the boundary between the shaded and unshaded regions. The idea that even simple mechanical systems such as a pendulum can behave chaotically is a fairly recent discovery. Chaotic behavior can usually be traced to the existence of nonlinear terms in the equation of motion of the system.

# Newton's Laws

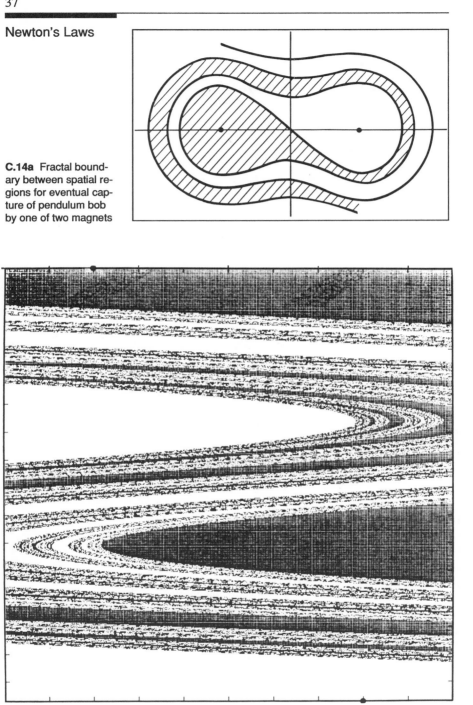

**C.14a** Fractal boundary between spatial regions for eventual capture of pendulum bob by one of two magnets

**C.14b** Magnified view of boundary showing its fractal character

# Center of Mass, Stability, and Friction

## D.1. Stacking metersticks

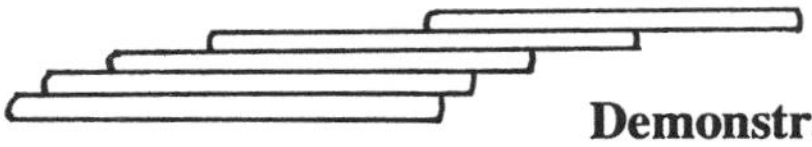

**Demonstration**

Metersticks can be stacked to give a surprisingly large projection of the top stick relative to the bottom one. The projection for a given number of sticks is a maximum when the center of mass of the top $N$ sticks lies directly above the end of the $(N+1)$st stick for all values of $N$.

**Equipment**

A collection of metersticks of equal mass. Remember that metersticks are not produced to have a common mass! (When I initially tried the demonstration using wooden sticks, I found that, depending on the type of wood, sticks that looked identical varied in mass by up to 100 percent.)

**Comment**

As shown below, the $N$th stick from the top can have a maximum projection relative to the stick below of $1/(2N)$ meters. For the 1st, 2nd, 3rd, 4th, 5th, and 6th metersticks, this formula yields maximum extensions of 0.500, 0.250, 0.167, 0.125, 0.100, and 0.083 meters, respectively, for a total extension of 1.225 meters for the first six sticks.

To prove the $1/(2N)$ formula we assume that the $N$th stick projects out a distance $d_N$ farther than the one below. We further assume that if the $N-1$ sticks above are just barely stable, their center of mass must lie directly above the edge of the $N$th stick. If we compute torques about a point at the edge of the $(N+1)$st stick, equilibrium requires that the torque due to the weight of the $N$th stick cancel the torque due to the weight of the top $N-1$ sticks. Equating the two torques, $(0.5 - d_N)mg = d_N(N-1)mg$, yields the $1/(2N)$ result for $d_N$. Based on this formula, the total extension of the first $N$ metersticks can be made arbitrarily large by choosing a sufficiently large $N$, since the sum: $\frac{1}{2} + \frac{1}{4} + \cdots + 1/(2N)$ is divergent. Although the sum diverges, it does so quite slowly, so that an exceedingly large number of metersticks is required to achieve a large extension of the top stick relative to the bottom one. For example, $1.5 \times 10^{44}$ metersticks would be needed to get an extension of 10 meters.

In carrying out the demonstration using seven metersticks you will probably want to give each stick an extension 1 cm less than the values predicted from the $1/(2N)$ formula, to allow for sticks of nonuniform weight. This leads to the arrangement shown in Plate D.1a for the first seven metersticks. The numbers indicated on each stick are the centimeter marks that should be made to line up. One way to rapidly assemble this pile stably is to predrill holes through sticks 2 through 7 at the centimeter marks indicated above, and then line up the holes with a rod placed through them. Place the top stick as indicated with its edge at the 49-cm mark of stick 2 and then slowly remove the rod.

A clever variation of this demonstration involves constructing a mobile using metersticks suspended by strings, as described in the article by Iain MacInnes, "An Instructive Mobile," *The Physics Teacher* 27, 42–43 (January 1989).

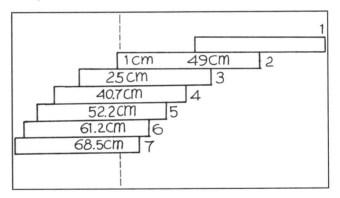

**D.1a** Lineup of seven metersticks for a near-maximum projection of the top stick

## D.2. Stability of a floating object

**Demonstration**

A floating object is in stable equilibrium only if its center of mass lies below its center of buoyancy.

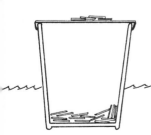

**Equipment**

A tray or dish 4 or more inches deep; a Styrofoam cup with a lid; and 50 pennies.

**Comment**

A floating object is acted on by two forces: its weight, which acts at the center of mass, and a buoyant force, which acts at the center of buoyancy, located at the geometric center of the submerged volume. The equilibrium of a floating object is stable if and only if slight disturbances of the object cause these two forces to produce a restoring torque driving the object back to the equilibrium orientation, which is the case only if the center of mass lies below the center of buoyancy. If the object is just barely stable, the centers of mass and buoyancy nearly coincide.

We can test this condition for stability using a floating Styrofoam cup. The cup can float if you place it in the water either right-side up or upside down, but the upside down orientation is the more stable one, because of the cup's sloping sides. For example, with the cup floating upside down you can probably add 2 or 3 pennies to the top before the cup tips over, while if it is floating right-side up you probably cannot add a single penny on top of the cover without causing the cup to tip.

In order to show that stability requires that the center of mass lie below the center of buoyancy, float the cup right-side up with 20 pennies placed inside, and put the lid on. Add pennies on the top of the lid one by one, thereby gradually raising the center of mass and reducing the degree of stability. Observe the maximum number of pennies, $N$, the lid can hold before the cup topples over, and also the distance $x$ the lid lies above the water line when the last penny is added. You must add the pennies carefully by distributing their weight evenly, especially as you add the last few pennies. We shall show that the number of pennies that can be added without tipping satisfies the relation $N = Mx/(x + 2y)$, where $M$ is the number of pennies in the cup, and $y$ is the amount of cup below the water line. ($y$ is the height of the cup minus $x$, the observed distance above the water line.) After you see how well this relation is satisfied using 20 pennies in the cup you might want to test it for several other values.

To prove the above formula, we note that the center of mass of the pennies at the top and bottom of the cup must lie a distance above the bottom of the cup given by $N(x + y)/(N + M)$. For the cup to be barely stable, this distance must also equal the height of the center of buoyancy above the bottom, $y/2$. Thus, ignoring the weight of the cup, we can solve for $N$ to obtain $N = Mx/(x + 2y)$, as stated previously.

## D.3. Pulling a sliding brick

### Demonstration
By pulling a sliding brick with a scale, you can show that the friction force depends on the normal force and the roughness of the surfaces, but not on the (constant) speed or the surface area of contact.

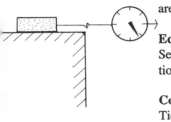

### Equipment
Several bricks; a board; tape; string; and a large demonstration scale.

### Comment
Tie or tape the string onto a brick, and attach the other end to the demonstration scale, which can be used to pull the brick across a board with a known force. With the string kept horizontal, slowly increase your pull on the scale until the brick begins to move. The scale reading should rise to some maximum value before sliding begins. You may notice a small, sudden drop once the brick starts sliding, showing that the force of sliding friction is less than the maximum force of static friction. You can easily observe that the force of sliding friction will be nearly independent of the brick's speed, as long as it is kept constant. The force is also independent of which brick face is in contact with the desk, showing that the surface area of contact doesn't matter. Add a second brick on top of the first one, and you will find that the force of sliding friction doubles, thereby verifying that the friction force is proportional to the force pressing the two surfaces together, known as the normal force. You can easily determine $\mu$, the coefficient of sliding friction, from the ratio of the friction force to the normal force, which here equals the weight of a brick (or two bricks).

An object that has a measured coefficient of sliding friction $\mu$ should slide down an incline at constant speed only if the angle of the incline is given by $A = \tan^{-1}\mu$. For other angles, a net force exists along the incline, and the brick must accelerate according to Newton's second law. You can test this by raising one end of the board to make an angle $A$ with the horizontal. (This can be done without a protractor, if you place a meterstick next to the raised end of the inclined board, and adjust its height to be $L \sin A$, where $L$ is the board's length.) The brick should not slide if placed at rest on the incline, since the coefficient of static friction exceeds the coefficient of sliding friction. But the brick should slide down the incline with constant velocity if you give it a little push. If you increase the angle of the incline, the brick should accelerate, and if you decrease it the brick should decelerate.

## D.4. Launching a sliding block with known velocity

**Demonstration**
By launching a sliding block with various initial velocities on a horizontal surface, you can observe the dependence of its stopping distance on the block's initial velocity and the coefficient of friction.

**Equipment**
A small wooden block; an eye hook; string; tape; sandpaper; and a metronome or a clock.

**Comment**
Screw the eye hook into the block and attach a few feet of string to it. You can launch the block across the floor with known velocity by whirling the block in a circle at a steady frequency and suddenly releasing the string. The tangential speed of the block can easily be found from the length of the string, $r$, and the rotation period $T$, which can be chosen to be one second if you synchronize the rotations with the clicks of a metronome or someone counting off the seconds. The block's speed at the instant you let go of the string is given by $v = 2\pi r/T$. If the coefficient of friction between the block and the floor is $\mu$, then by Newton's second law the deceleration of the block is given by $a = F/m = \mu mg/m = \mu g$. The predicted stopping distance $s$ is then given by $s = v^2/2a = v^2/2\mu g$. Thus for a floor of constant roughness you should find that the stopping distance varies as the square of the launch speed.

The easy way to test this relationship is to vary the launch speed by varying the length of string while keeping the rotation period constant at one second. The radius of the circle equals the length of the string plus half the width of the block. Be sure to keep the block in continuous contact with the floor as you whirl it in a circle. You can verify the $v^2$ dependence of the stopping distance by seeing whether a plot of $v^2$ versus $s$ yields a straight line. You should also test how the stopping distance depends on the mass of the block, and on the surface roughness. The mass can be varied by taping two blocks together, and the surface roughness can be varied by adding some masking tape or sandpaper to the underside of the block.

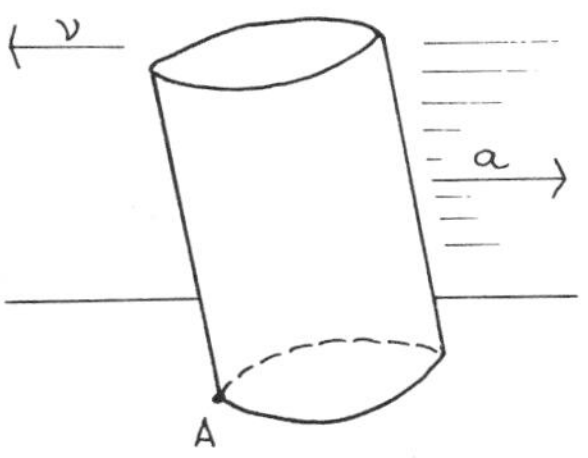

## D.5. No tipping allowed

**Demonstration**
A cylinder slides to a stop on a horizontal surface without tipping only if its height-to-diameter ratio is less than the reciprocal of the coefficient of kinetic friction.

**Equipment**
An assortment of cylinders of equal diameter that have a range of heights; and a smooth, wide board that can be inclined at various angles. The cylinders could be cut from a broom handle, but if you are giving the demonstration to a large group, the cylinders should have a large radius. Segments of a 3-in.-diameter polyvinyl chloride (PVC) pipe would be good if you can cut them so that they have a base that is flat and perpendicular to the lateral dimension.

**Construction**
As shown below, the height of the tallest cylinder that can be brought to rest without tipping while sliding on a horizontal surface is given by $H = d/\mu_k$, where $d$ is the cylinder's diameter and $\mu_k$ is the coefficient of kinetic friction. You should cut a collection of cylinders of common diameter $d$, whose heights $h$ bracket the value $H$, such as the set having $h/H = 0.75, 0.85, 0.95, 1.05, 1.15$. You need to know the value of $\mu_k$, of course, before you can cut the cylinders to the proper heights. $\mu_k$ can easily be determined by placing a short cylinder on an inclined board, and finding the angle of the incline for which the cylinder slides at constant speed after being given an initial push. The coefficient of kinetic friction is given by $\mu_k = \tan A$, where $A$ is the angle of the incline.

**Comment**
Observe which of the cylinders can occasionally be brought to rest without tipping when given an initial push on a horizontal surface. (Push them on the same smooth board used to find $\mu_k$.) If $\mu_k$ has been determined correctly, you should find that the cylinders of heights $0.85H$, $0.95H$, and $1.05H$

never tip, sometimes tip, and always tip, respectively. For best results use a block of wood to apply a uniform pressure on the bottom half of the cylinders, and resist the temptation to give them an extra push just as they leave the block.

In order to derive the condition $H = d/\mu_k$, we note that when the cylinder is on the verge of tipping, the upward normal force $N$ acts at the leading edge of the base (point $A$ in the illustration). In the decelerating reference frame moving with the cylinder, a horizontal inertial force, $-ma$, acts at the center of mass, along with a vertical gravitational force, $mg$. The maximum height of a cylinder for which no tipping occurs can be found by requiring that the resultant of these forces be directed toward point $A$, i.e., along an angle with the vertical given by $\tan^{-1}(d/H)$. Thus, we require that (1) $a/g = d/H$, if the cylinder is on the verge of tipping. In addition, we know that for an object sliding on a surface with coefficient of kinetic friction $\mu_k$, the object's acceleration is given by (2) $a = \mu_k g$. Combining equations (1) and (2), we see that the maximum height for no tipping is given by $H = d/\mu_k$.

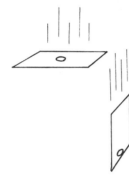

## D.6. Air resistance as a form of friction

### Demonstration
The dependence of air resistance on the shape, orientation, density, size, and velocity of an object can be shown by dropping pairs of objects from a common height.

### Equipment
A napkin; a textbook; two Styrofoam balls, 1 inch and 3 inches in diameter, obtainable from a crafts store; two rubber balls, 1 inch and 3 inches in diameter; several index cards; and pennies taped to each card. One index card should have a penny taped to its center, and the other a penny taped to the middle of a short side's edge.

### Comment
Air resistance is a major problem for designers of cars, airplanes, and rockets. Of course, without air resistance many other devices, including parachutes and rocket reentry vehicles, would be impossible. In the absence of air resistance a very light object, such as a feather, falls with the same acceleration as a heavier object, such as a coin, as can be verified by dropping them together in an evacuated tube.

**Stability and Friction**

A much simpler way of demonstrating that light and heavy objects fall with different accelerations only because of air resistance can be shown using a paper napkin and a textbook. If the napkin is placed on top of the book and the two are dropped, they fall together as long as the book remains horizontal and shields the napkin from the force of the onrushing air. If the book and napkin are dropped side by side, the effect of air resistance is much more pronounced on the napkin.

The force of air resistance on an object can be expressed as $CAv^2$, where $A$ is the object's surface area, $v$ is its velocity, and $C$ is a constant that depends on its shape and orientation. According to Newton's second law, the acceleration of a falling object is $F/m$, where the net force $F$ is the difference between the upward force of air resistance and the downward weight $mg$, so that $a = F/m = CAv^2/m - g$. In the case of a falling sphere, if we substitute volume times density ($Vd$) for the mass $m$, and use $R/3$ (where $R$ is the radius of the sphere) for the ratio of a sphere's volume to its surface area, $V/A$, we obtain $a = 3Cv^2/Rd - g$. Based on this equation, air resistance therefore causes spheres to descend with an acceleration less than $g$ whenever $3Cv^2/Rd$ is not negligible, or whenever the velocity is sufficiently large, the radius sufficiently small, or the density sufficiently small. The dependence of air resistance on these three factors (as well as on the constant $C$, determined by the object's shape and orientation) can easily be observed by dropping pairs of objects.

1. Shape. A crumpled piece of paper falls faster than an uncrumpled piece, showing the importance of shape.
2. Orientation. If the two index cards with pennies taped on them are dropped from a common height with a horizontal orientation, the one with the penny near an edge quickly assumes a vertical orientation, and it falls faster than the other one, which remains horizontal.
3. Density and size. Large rubber and Styrofoam balls fall together, but small rubber and Styrofoam balls do not, indicating that an object's density becomes important when it is sufficiently small in diameter, or conversely that an object's size becomes important when its density is sufficiently low. For a large, slowly moving object, air resistance is too small to be observable unless its density is extremely low, as in the case of a balloon. These observations show that the importance of air resistance depends on the product $Rd$.

**4.** Velocity. Small Styrofoam and rubber balls fall together when dropped from a height of 3 feet, but not when dropped from a height of 6 feet. A Ping-Pong ball and a rubber ball fall together when dropped from 6 feet, but not 12 feet, where higher velocities are attained. (You will probably need to stand on a desk to do this part of the demonstration.) Large and small rubber or metal balls fall together unless they fall a distance much greater than 12 feet. These observations are explained by the $v^2/Rd$ dependence of the force of air resistance, since the balls must acquire appreciable velocity in order for air resistance to become important—the more so the higher their densities $d$ and the larger their size.

The same explanation can be given for the fact that a dropped object and a horizontally projected object hit the ground simultaneously, but only if they have high density or low velocity (see demonstration A.1). For a pair of low-density or high-speed objects, the one that is horizontally projected experiences the greater air resistance owing to its higher velocity, and it reaches the floor after the dropped object. The effects of air resistance on dense objects can be quite important at very high speeds. For example, an M1 rifle bullet, which leaves the gun at about 990 m/s, has a maximum range that is only 2.5 percent of its predicted range ignoring air resistance.

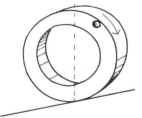

## D.7. Weighted Styrofoam sphere and disk on an incline

### Demonstration
The dependence of stability on initial conditions is shown by the fact that weighted spheres and disks can be made to either roll up or down an incline, or remain at rest, depending on their initial orientation.

### Equipment
An 18-in. Styrofoam disk and a 4-in. Styrofoam ball (obtainable at a crafts store); two lead or steel balls; a hacksaw; plastic cement; and duct tape.

### Construction
Saw the Styrofoam sphere in half with a hacksaw, and create two cavities in each half by pressing a small lead or steel sphere into the Styrofoam near the edge. The spherical cav-

ity should be just large enough to hold the metal sphere. Using plastic cement, glue the hemispheres back together with the metal sphere in place, and mark the spot on the Styrofoam sphere directly above the metal sphere inside. If you are worried about the metal sphere getting thrown free if the hemispheres break apart, use duct tape to seal the seam between the hemispheres. To make the weighted disk, hollow out a small cavity at some point at the center of the edge of an 18-in. Styrofoam disk. Ensure that the cavity is not any larger than a lead or steel ball by pressing the ball into place. Wrap duct tape around the circumference of the disk to keep the ball in place. A sufficiently heavy metal ball should be used so that its weight is greater than that of the disk.

**Comment**

If the sphere is placed on an incline of less than some critical angle, it will not roll down provided its initial orientation puts the metal ball near the contact point with the incline. The critical angle depends on both the mass ratio and the radius ratio of the metal ball to the Styrofoam sphere. It can be shown, using the second condition for equilibrium, that if the radius of the metal ball is small compared to that of the sphere, the critical angle is given by $A = \sin^{-1}[1/(1 + m/M)]$, where $m/M$ is the ratio of the mass of the Styrofoam sphere to that of the metal ball. Clearly, if the mass ratio $m/M$ is small compared to 1.0, the critical angle can be very large— so large, in fact, that the sphere may actually *slide* down the incline before the critical angle for rolling is reached. Thus, if you experimentally test the preceding equation for the critical angle, and you find that as you increase the angle of the incline, sliding occurs before rolling, you will need to use a less massive ball or else wrap the sphere with tape to increase the coefficient of friction.

When the sphere is in equilibrium on an incline of less than the critical angle, the center of mass of the sphere plus ball must lie directly above the point of contact. The equilibrium is stable because small rotations of the sphere in either direction would raise the center of mass. However, this is not true of large rotations, so if the sphere's initial orientation puts the metal ball far in front of the contact point (the down-incline side), the sphere rolls down the incline even if its angle is small. If the sphere's initial orientation puts the metal ball far behind the contact point (the up-incline side), the sphere begins to roll up hill, possibly all the way up if it is placed close to the top. The "gravity-defying" behavior of the disk is even more impressive than that of the sphere, because its larger radius allows it to roll farther uphill. If the incline is too steep, however, the weighted disk may flop over rather than roll up.

## D.8. Pennies balanced on a ruler

**Demonstration**
Balancing a ruler with pennies on it illustrates the equality of clockwise and counterclockwise moments (torques) about its center—the second condition for equilibrium.

**Equipment**
A plastic ruler; some pennies; and a fulcrum (such as a pencil). A small triangular block would make a better fulcrum than a pencil, since it would allow larger swings of the ruler—an important consideration if the demonstration is done on an overhead projector.

**Comment**
Place the plastic ruler on the fulcrum on an overhead projector and balance the ruler at its center. Place one penny on one side of the fulcrum and two pennies on the other side. The two pennies should be next to each other rather than stacked, so that they can clearly be seen as two. Balance will be attained when you make the single penny's distance to the fulcrum twice the distance of the center of mass of the two pennies to the fulcrum. You can repeat the demonstration using a number of pennies at various locations on each side, and show that for balance to be achieved, the sums of the clockwise and counterclockwise moments must be equal. Doing the demonstration using a transparent ruler makes it easy for a group to see the distances to the fulcrum for each penny. Since you don't want to worry about the moment for the ruler's weight, be sure to keep its center at the fulcrum. However, if you first determine the ruler's weight in pennies by sliding the ruler over the fulcrum until its weight acting at the center of gravity balances a single penny placed at the end of the ruler, you can place the fulcrum anywhere, as long as you include the moment acting at the ruler's center of gravity in the moment equation.

The center of gravity of the ruler lies at its center only because of its uniform cross-sectional area. An interesting variation is to replace the ruler with a "tapered rod" of nonuniform cross section, made from a piece of thick cardboard cut into a triangular shape with a base of 1 inch and an altitude of 8 inches. It can easily be shown that the center of gravity of this tapered rod is located at a point one-third the length from the thick end, a fact that can be verified by seeing where the rod balances.

## D.9. Moving two fingers under a meterstick

**Demonstration**
The relationship between the coefficients of static and kinetic friction explains why two fingers supporting the ends of a meterstick always meet at the center of the stick as they are brought together, even if the fingers are initially located at different distances from the center.

**Equipment**
A meterstick.

**Comment**
According to the second condition for equilibrium, the fractions of the meterstick's weight resting on your two fingers, $W_1$ and $W_2$, depend on the distances $x_1$ and $x_2$ to the center, according to the relation $W_1x_1 = W_2x_2$. Just at the point where one finger stops moving and the other starts moving, the static-friction force of the fixed finger equals the kinetic-friction force of the moving finger: $\mu_s W_1 = \mu_k W_2$. Combining the two preceding equations yields the condition $\mu_k x_1 = \mu_s x_2$. By observing just where one finger stops sliding and the other starts, you can measure the values of $x_1$ and $x_2$, and thereby determine the ratio of the two friction coefficients using $\mu_k/\mu_s = x_2/x_1$.

The best procedure for measuring the ratio of the two friction coefficients is to start with the meterstick balanced on your two fingers with one finger (say, the left) near the very end of the stick and your right finger 10 cm in from the end (a distance $x_1 = 40$ cm from the center). As you bring your fingers together, your left one will initially slide until it reaches a point (a distance $x_2$ from the center) where it comes to rest and the right finger starts sliding. As already noted, the ratio of the kinetic and static friction coefficients can be found from the measured ratio $x_2/x_1$, which you should find to be about 0.5.

But why must your fingers always meet at the center of the stick as the sliding keeps switching between left and right fingers? The ratio of the distances to the center from the points at which each switch occurs must equal the ratio of the two friction coefficients, which is a constant. Accordingly, the differences between successive distances $x_1$ and $x_2$ get progressively smaller as your fingers approach the center. Theoretically, an infinite number of switches should occur between the two fingers' motions before they meet at the center.

You can, of course, make your fingers meet at some point other than the 50-cm mark by suddenly accelerating one of them, because in that case, the friction force each finger exerts on the stick need not be the same. According to Newton's second law, accelerations imply unbalanced forces. Another variation on the demonstration would be to hang unequal masses at the ends of the meterstick, and locate their center of mass by sliding your fingers under the stick. (This actually locates the center of mass of the stick plus the masses, which should be selected to have a large mass compared to that of the stick.)

# Energy and Linear Momentum Conservation

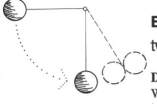

## E.1. Inelastic collisions between two balls

### Demonstration
When one Velcro-covered ball is swung through a 90° arc into a second ball, conservation of momentum requires that the center of mass of the two balls reach one-fourth the initial height of the first ball.

### Equipment
Two balls the size of tennis balls or larger; some Velcro with adhesive backing, obtainable at a crafts or fabric store; and string.

### Comment
The two balls should be entirely covered with Velcro, and tied to the two ends of a string held at its middle. When one ball is released from a 90° angle and allowed to strike the other, momentum conservation requires that the two balls together have half the velocity of the first ball just before impact. Energy conservation then requires that the center of mass of the two balls rise to a height of one-quarter the first ball's height above the lowest part of the swing. In order to verify this prediction, the half-length of the string, $L$, should be large compared to the size of the balls. But if you use a string longer than about 2 or 3 feet it will be difficult to achieve head-on collisions. You could, however, suspend each ball trapeze-style using two strings attached to a bar in order to facilitate head-on collisions. Rather than directly observing if the balls rise to a height $L/4$ above the lowest part of the swing, it may be easier for you to observe if they swing to the correct final angle $A$, where $A = \cos^{-1} 0.75 = 41°$, measured from the vertical. This can be tested most easily by swinging the balls in front of a blackboard or overhead-projector screen on which you have marked a line making a 41° angle with the vertical.

# E.2. Rolling disks, hoops, and spheres down an incline

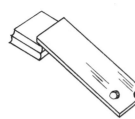

### Demonstration

Energy conservation requires that when disks, hoops, and spheres are rolled down an incline, the order in which they reach the bottom is spheres, disks, and hoops—regardless of their radii or masses, or the slope of the incline.

### Equipment

Disks, hoops, and spheres of two different radii each; and a board at least 3 feet long propped up by some books to serve as an incline. The disks and hoops should be wide enough to not topple over while rolling. The board should be wide enough to accommodate several objects rolling side by side. Be sure that the spheres are not hollow. It would also be interesting to include a sealed can one-third full of sand. An alternative method of display suited for use on an overhead projector uses a half-inch-thick piece of acrylic sheet inclined at a very slight angle, in lieu of a board. The advantage of the slight angle is that objects rolling down the incline remain in focus the entire time, and the differences in roll times for different shapes are greater. If the angle is too small, however, the smaller rolling friction of a sphere compared to a disk or a hoop introduces spurious results.

### Comment

An object of radius $R$ rolling down an incline acquires rotational and translational kinetic energy as it loses potential energy. It can easily be shown that the ratio of the rotational and translational kinetic energies is given by $(k/R)^2$, where $k$ is the radius of gyration that appears in the general formula for an object's moment of inertia about its center of mass, $mk^2$. Conservation of energy, therefore, requires that the speed of an object at the bottom of an incline be given by $v^2 = 2gh/[(k/R)^2 + 1]$, independent of its mass $m$ or radius $R$. Objects that have a small $k/R$ ratio (more mass close to the center) yield a larger velocity $v$, because a smaller fraction of the original potential energy is diverted to rotational kinetic energy, and more energy is available for energy of translation.

The values of $k/R$ for the sphere, disk, and hoop are $2/5$, $1/2$, and 1, respectively. Of the three objects, the sphere with the smallest $k/R$ ratio has the highest velocity, and the hoop with the largest $k/R$ ratio has the lowest velocity. The order of descent is therefore sphere, disk, and hoop, regardless of radius or mass, which can easily be verified using pairs of rolling objects, and observing which one reaches the bottom first.

We can make such "races" quantitative by giving one object of each pair the head start necessary for them to reach the bottom simultaneously. If the times of descent are the same for two rolling objects, the ratio of the distances they travel must equal the inverse ratio of their final velocities. For example, in a race between a disk and a hoop, they reach the bottom simultaneously if the ratio of the distances traveled by the hoop and the disk is $(0.5^2 + 1)/(1^2 + 1) = \frac{5}{8}$. In other words, the slower hoop must be given a head start of $\frac{3}{8}$ (37.5 percent) the length of the incline for it to tie the disk at the finish line. Similarly, in a race between a sphere and a hoop, the slower hoop must get a 42 percent head start for the race to be a tie.

These predictions can easily be tested if you release the hoops, spheres, and disks from predrawn starting lines appropriate to each. For best results keep the angle of the incline small in order to obtain longer times of descent. But remember that friction and the flatness of the incline assume greater importance the smaller you make the angle.

We have so far assumed that mechanical energy is conserved during the descent. If, however, a can one-third full of sand or lead shot is rolled down the incline, an appreciable fraction of the original potential energy is converted to heat as the sand is tossed about during the can's descent. Accordingly, less energy is available for linear motion down the incline, and the can descends more slowly than the three rigid objects. In fact, you may find that the can doesn't roll at all if the incline is sufficiently gentle, as explained in demonstration D.7.

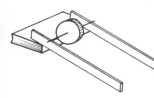

## E.3. Rolling a wheel with axle down a slight incline

### Demonstration

Energy conservation requires that when a wheel connected to an axle rolls down an incline on the axle, nearly all of its kinetic energy be rotational, so it descends very slowly. Likewise, the descent of the wheel and axle is also very slow when it is suspended by a pair of strings wound around the axle and released.

**Equipment**

Two metersticks; and a wheel connected to an axle with a small diameter.

**Construction**

Make an 8-in.-diameter wheel out of a piece of plywood or acrylic plastic sheet. (You could use the disk made for demonstration C.4.) It need not be perfectly round as long as you drill out its exact center of gravity to mount the axle. (You can locate its center of gravity by seeing where it balances on a nail.) Make the axle out of an ⅛-in.-diameter threaded rod or a long machine screw. Be sure that the axle fits loosely through the hole and extends at least an inch beyond the wheel on either side. Firmly tighten nuts and washers on both sides of the wheel to rigidly connect wheel and axle. The incline down which the wheel and axle is rolled is created by resting the two ends of the metersticks on two stacks of books, one of which is 6 inches higher than the other. The sticks should be edge up, not face up, so that they don't flex when the axle is placed on them.

**Comment**

Rest the axle on the two metersticks and allow it to roll down the incline. (You may need to give it a little push.) The acceleration down the incline is surprisingly small, due to the small magnitude of the diameter of the axle relative to that of the wheel. You can easily time how long it takes to roll 10, 20, 30, 40, . . . centimeters, and see if the acceleration is uniform.

At the bottom of the incline the initial potential energy $mgh$ has been converted entirely into translational and rotational kinetic energy. The rotational kinetic energy, $\frac{1}{2}I\omega^2$, can be computed using $I = \frac{1}{2}MR^2$ and $\omega = v/r$ to obtain the relation $K_{rot} = \frac{1}{4}mv^2(R/r)^2$, where $R$ is the disk radius and $r$ is the axle radius. From the preceding expression for the rotational kinetic energy, we see that the rotational kinetic energy is larger than the translational kinetic energy by the factor $\frac{1}{2}(R/r)^2$. For the case of an 8-in.-diameter wheel and ⅛-in.-diameter axle this factor becomes $\frac{1}{2}(64)^2 = 2048$, which explains why the wheel rolls down the incline with so little translational (and so much rotational) kinetic energy.

The same factor of 2048 for the ratio of rotational and translational kinetic energy holds when the wheel and axle are suspended by a pair of strings wound around the axle and tied to a horizontal bar. Assuming the thickness of the strings does not lead to a larger effective axle radius, the vertical acceleration of the wheel should be given by $g/(2048)^{1/2}$, or 0.22 m/s². With this acceleration, the wheel would take as long as 3.0 seconds to descend one meter. If, however, you loosen the nuts connecting the wheel and axle, the wheel descends with nearly 9.8 m/s² acceleration, since it acquires very little rotational kinetic energy in its descent, and nearly all the original potential energy can go into translational kinetic energy of the falling wheel. (This part of the demonstration does not work on the incline owing to the friction between wheel and axle.)

## E.4. Collisions on a vibrating plate

### Demonstration
By observing the dot trails left by balls colliding on a vibrating plate covered with carbon paper, you can find the balls' vector velocities, and test momentum and energy conservation in two-dimensional collisions.

### Equipment
See demonstration A.4.

### Comment
As explained in demonstration A.4, balls rolling on a vibrating plate covered by face-up sheets of carbon and plain paper (or a transparency film) leave trails of dots spaced by $1/60$ sec owing to the pressure fluctuations. If one ball is placed at the center of the paper and a second ball is rolled toward it, dot trails similar to those in Plate E.4a will be created.

In Plate E.4a, a projectile ball was rolled toward the right, along trail $A$. After it bounced off a stationary target ball, it rolled away along trail $B$. The stubby line ($C$) at the beginning of trail $D$ left by the outgoing target ball shows that the target ball was not quite stationary. The curvature of trail $B$ is due to the mismatch between the directions of the linear and rotational velocities of the projectile ball immediately after the collision.

Since we seek the velocities of the balls immediately after the collision, we must draw a velocity vector *tangent* to track $B$ at the point of collision. The length of the velocity vector is then made equal to the distance between some spe-

## Section E

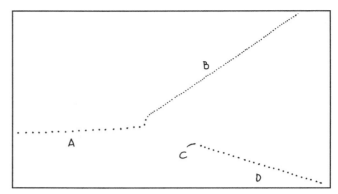

**E.4a** Dot trails created by colliding balls on a vibrating plate

cific number of dots corresponding to a particular time interval $dt$. The same number of dots must of course be used to construct all three vectors. You will want to use enough dots to average out irregularities in their spacing, but not so many dots that you include regions of the dot trails where the ball has clearly changed its velocity.

An important complication arises in drawing the velocity vector for the target ball after the collision (dot trail $D$ in Plate E.4a). The measured velocity of this ball must be multiplied by the correction factor $7/5$ to find its velocity immediately after the collision. This correction is necessary because the target ball immediately after collision is given a linear velocity, but no spin, and is therefore sliding rather than rolling. In a very short time the force of sliding friction slows the ball to $5/7$ its initial speed, and simultaneously causes it to have enough spin to meet the condition necessary to achieve rolling.

If you draw all three velocity vectors to scale on graph paper, then, on the assumption that $\vec{V}_C = 0$, you should find that $\vec{V}_A = \vec{V}_B + \vec{V}_D$, as required by momentum conservation for balls of equal mass. In other words, the three vectors $-\vec{V}_A$, $\vec{V}_B$, and $\vec{V}_D$ should form a closed triangle. You can also test the extent to which energy is conserved in the collision. For balls of equal mass, conservation of energy requires that $V_A^2 = V_B^2 + V_D^2$. [An even simpler check would be to see if the two ball directions after collision ($\vec{V}_B$ and $\vec{V}_D$) are at right angles.]

Remember that you need to draw a tangent to dot trail $B$ at its point of origin to get its true direction immediately after impact. Best results will be obtained for those cases in which $\vec{V}_C$ is most nearly zero—no short stubby track for the target ball, which is assumed to be stationary.

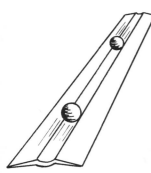

## E.5. Loop-the-loop on an incline

**Demonstration**

According to the law of the conservation of energy, a ball rolled down a loop-the-loop track resting on a gently inclined flat surface completes the loop-the-loop only when released above a certain height.

**Equipment**

Shape a "flexible curve" (a drawing device obtainable from an art supply store) into the contour shown above. The flexible curve should be at least 16 inches long. When the curve is placed on a gently inclined surface a steel ball can be rolled down the curve.

**Comment**

The ball will remain in contact with the curve at the top of the circle only when released from a point whose height $h$ above the bottom of the circle exceeds 2.7 times the loop radius $r$—a result that is independent of the angle of the inclined surface. The effect of releasing the ball at a variety of heights can be easily seen on an overhead projector if the inclined surface is transparent. If no such transparent material is available you can just prop up one edge of the projector itself. The condition that $h$ exceed $2.7r$ follows from conservation of energy, and the requirement that at the top of the circle $mv^2/r$ be no less than $mg$. A more detailed analysis shows that the condition that $h$ exceed $2.7r$ is only an approximation, whose accuracy depends on the coefficients of static and kinetic friction. Note that if we had neglected the rotational kinetic energy of the ball, we would have obtained $2.5r$ for the minimum height $h$, which applies in the case of a frictionless sliding mass.

## E.6. Colliding balls on a grooved ruler

**Demonstration**

When balls are rolled on a grooved ruler, you can observe the kinds of collisions that occur between balls having different masses and elasticities.

**Equipment**

A grooved, clear plastic ruler that has the cross section shown in the illustration; and an assortment of small balls no bigger than 1.5 inches in diameter. The balls should be made of materials such as metal, wood, or plastic (not rubber), which allow them to slide rather than roll when projected at high speed. Steel balls are particularly good. The plastic ruler needs to be transparent only if you want to show the demonstration on an overhead projector.

**Comment**

First try rolling two balls at roughly equal slow speeds toward each other in the ruler groove. (It is surprisingly easy to make the two speeds roughly the same by moving your two hands in mirror image motions.) You may find that the balls lose some of their energy during the collision even though they may be made of a highly elastic material. Energy is lost because the collision, which reverses the balls' linear velocities, does not reverse their spin directions. The balls, therefore, have backspin after the collision, which causes their speeds to slow rapidly until the linear and rotational speeds are properly matched for rolling motion. The energy loss appears to be proportionately less for high-speed collisions, possibly because by rapidly sweeping your hands together you launch the balls with a motion that is more sliding than rolling.

One way to demonstrate that translational kinetic energy is more nearly conserved when the balls' speeds are high is to launch one ball ($A$) into a stationary ball of equal mass ($B$). If energy were exactly conserved, ball $A$ should be at rest after the collision. You will probably find that after the collision, ball $B$ has proportionately less speed, the higher the initial speed of ball $A$. In addition to observing one ball colliding with a target ball of equal mass, you should also observe the following:

(1) Elastic collision between two balls of equal mass having equal and opposite velocities.

(2) Elastic collision between one ball and a row of two or more stationary balls of equal mass.

(3) Elastic collision between a (light) Ping-Pong ball and a (heavy) steel ball moving toward each other with equal speeds. You should find that the Ping-Pong ball rebounds with a much increased speed (ideally, three times its original speed if spin is ignored), and the steel ball's speed is essentially unchanged by the collision.

(4) Elastic collision between two metal balls whose mass ratio is 3 : 1, moving toward each other with equal speeds. If momentum and energy are both conserved, and spin is

ignored, the heavy ball should be at rest after the collision, and the light ball should double its speed. In practice, the spin of the heavy ball results in its possessing some forward motion after the collision.

(5) Inelastic collision between two balls of equal mass with equal and opposite velocities. For the inelastic collision you can use two "no-bounce" balls (see A.1). Ping-Pong balls sawed in half, filled with clay, and glued back together are also good.

When observing collisions between balls of very different mass, it is best to use balls of roughly the same size and different densities, rather than different sizes and the same density, so the balls collide nearly center to center.

## E.7. Momentum conservation using an embroidery hoop

### Demonstration
When two touching, stationary balls inside a horizontal embroidery hoop are driven apart with a sharp vertical blow, momentum conservation requires that the ratio of the angles they travel before colliding equal the inverse ratio of their masses.

### Equipment
A 10-in. embroidery hoop; several 1-in.-diameter balls of different densities (the 3 : 1 ratio of steel and aluminum works well); and an ordinary table knife, whose handle is used to deliver the vertical blow. If you are doing the demonstration on an overhead projector, you may wish to place an acrylic plastic sheet under the hoop to protect the glass surface of the projector when you strike the balls. A transparency showing angles marked every 5 degrees should be taped onto the acrylic sheet. A piece of polar-coordinate graph paper with appropriate lines darkened can be photocopied from demonstration C.4, and used to make the transparency.

### Comment
If the knife handle delivers its blow vertically, the ratio of the speeds of the balls must equal the inverse ratio of their masses, in order that the net momentum remain zero, as it was before the rod struck. The angular distances the balls

travel to reach their point of impact must likewise have the same ratio, because the balls each take the same time to reach the impact point.

To conduct the demonstration, place the balls, hoop, polar-coordinate transparency, and acrylic sheet on the overhead projector. The balls may wander if the projector is not exactly level, so you will need to either level the plastic sheet using index cards as shims, or else place the balls at that point of the hoop where they will stay put. Tape the hoop down to center it with respect to the polar coordinate transparency.

When you strike the balls with the knife handle, they receive equal and opposite momenta only if (a) you deliver the blow vertically, (b) you aim for the point of contact of the balls, and (c) the balls have the same radius, and negligible coefficient of sliding friction. Large differences between the ball's momenta can result if the blow is not delivered straight down and directly toward the contact point. With practice, however, it is easy to observe angles of the impact point all falling within a 20° arc on repeated trials. By averaging the angles from a large number of trials, the uncertainty in the average angle can be reduced to a few degrees.

Momentum conservation can be tested by seeing if the ratio of the angles traveled by each ball to the impact point equals the inverse ratio of their masses (found by weighing them). One small correction, however, needs to be made before computing the angle ratio. You must subtract the angular diameter of each ball from the angle it travels to reach the impact point, because you want to use the angle that the ball's center of mass moves from the impact point.

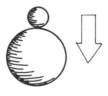

## E.8. Dropping a small ball on top of a big ball

### Demonstration
Conservation of momentum and energy requires that when a Super Ball of small mass resting on top of a Super Ball of large mass falls from a low height, the ball of small mass ideally rebounds to nearly nine times its original height.

### Equipment
One large Super Ball and one small Super Ball (or a Ping-Pong ball). A basketball would be a good alternative to the large Super Ball, because the small Super Ball would be less likely to roll off during the balls' descent.

**Energy and
Momentum
Conservation**

**Comment**

When the balls are dropped, both balls attain a downward velocity $v$ just before striking the floor. The bottom ball reverses its velocity an instant before the top one, so the small ball moving downward with speed $v$ strikes a large ball moving upward with speed $v$, making the two balls' *relative* speed $2v$. The bottom ball, being much more massive, can be thought of as a moving wall. Whenever a ball elastically bounces off a moving wall, the magnitude of its velocity relative to the wall stays the same, although its direction is reversed.

Before the top and bottom balls collide, their relative velocity is $2v$, so after impact their relative speed must still be $2v$. This will require the top (light) ball to have an upward speed of $3v$, because the bottom (heavy) ball continues with nearly its original speed $v$. Because the top ball's velocity is tripled as a result of the impact, its maximum height becomes nine times its original height, a consequence of the relation $v^2 = 2gy$. You can best test this prediction if the radius of the small ball is much less than that of the big ball, and they are dropped from a low height, so the small ball doesn't have much chance to roll off center during the fall.

An extension of the preceding demonstration is to drop a stack of three balls, with the top one much lighter than the middle one, and the bottom one much heavier—for example, a Ping-Pong ball on top of a Super Ball, on top of a basketball. It can be shown that for elastic collisions, the height of the Ping-Pong ball's rebound approaches a maximum of 49 times the initial release height. Although it is difficult to carry out this part of the demonstration, if you drop the balls from a height of a few inches, you will find that on occasion the three balls stay sufficiently aligned during their descent for the Ping-Pong ball to rebound to a great height. The chances of a favorable alignment can be considerably improved by placing the Ping-Pong ball and Super Ball inside a tube from a toilet-paper roll which rests on top of the basketball.

Still another variation of this demonstration uses two balls with a mass ratio of $3 : 1$, approximated by a baseball and a basketball. We assume, as before, that the basketball (on bottom) reverses its velocity a split second before the baseball reverses its velocity. Momentum and energy conservation then require that, after the collision between the balls, the basketball is at rest and the baseball has twice its speed before impact (causing it to reach four times its initial height). Thus, a baseball on top of a basketball dropped from a height of 3 feet should leave the basketball dead on the floor, and the baseball hitting the ceiling, assuming the collision is reasonably elastic.

## E.9. One-dimensional double-well potential

**Demonstration**
A one-dimensional double-well potential constructed from a flexible curve can be used to illustrate conservation of energy. A variation of the demonstration illustrates the transition from order to chaos.

**Equipment**
A flexible curve at least 16 inches long, obtainable at an art supply store; a metal ball; and a transparent sheet of acrylic or glass on which to place the flexible curve. (If you don't have a transparent sheet, you can simply put the curve on the overhead projector and tilt the projector a bit.) A related demonstration uses a supply of BB's and a piece of grooved rubber tread shaped into a double-well potential.

**Comment**
If friction is ignored, the sum of the kinetic and potential energy stays constant. Thus the ball moves faster the lower its height on the curve, a qualitative connection that can easily be observed when the ball is released from any point above the bottom of either well. When the ball is in the upper (left) well in the illustration, it oscillates in the well if it is given a speed less than that needed to reach the top of the well. When the ball is given enough energy to leave the left well, it can reach the right well, but may not have enough energy to get back over the hump, since some energy is lost because of friction.

An interesting related demonstration involves a large number of balls simultaneously rolling in identically shaped double-well potentials. This can be accomplished by rolling BB's in the parallel grooves of a piece of rubber tread that has been shaped into a symmetrical double well by placing the rubber tread into a frame to hold its shape (see Plate E.9a). In Plate E.9a the BB's, which are free to roll in each groove, are initially at rest at the bottom of the left well. If the apparatus is *gently* shaken back and forth, the BB's all oscillate together as long as the shaking continues. When the apparatus is shaken more vigorously, however, there is a

**E.9a** Apparatus to show the transition from order to chaos for balls rolling in a double-well potential

different outcome. Once the amplitude of shaking is sufficient for BB's to overcome the middle hump between wells, *chaos* ensues: BB's that have nearly the same initial conditions (being at rest at the bottom of one well) no longer remain together as time proceeds. This demonstration shows that the same system can exhibit either regular or chaotic behavior depending on the exact conditions—here, the amplitude of the driving oscillations. The capability of a system to exhibit either regular or chaotic behavior was also illustrated in demonstration C.14, the key determinant in that case being how close the initial release point of the pendulum bob was to the fractal boundary separating two regions of the plane where the swinging pendulum eventually came to rest next to one of two magnets.

# E.10. Energy storage in a rubber hemisphere

**Demonstration**
The transformation of elastic potential energy to kinetic energy is shown by the extreme height to which a hollow rubber hemisphere jumps when it is turned inside out and dropped from a low height.

**Equipment**

Half of a hollow rubber ball, 2¼ inches in diameter. (You can either cut a ball—the kind used for racquetball—in half, or buy the novelty item named "hopper popper" from Jerryco, Inc.)

**Comment**

Because of its initial curvature in manufacture, a hollow rubber hemisphere that is turned inside out stores a lot of energy. (You can easily see this by popping the hemisphere from one equilibrium configuration to the other, and seeing that one popping direction is a lot more vigorous than the other.) Energetically, this situation is very much like the case of the two potential wells in E.9. A ball in either well is in a stable state, but it gains (or loses) kinetic energy in going from the higher (or lower) well to the other, the gain in kinetic energy coming at the expense of a loss in potential energy.

One way to show the energy release is to pop the hemisphere into the inside-out configuration and drop it from a low height, flat-side down. Dropping the hemisphere is equivalent to giving the ball in the upper well of demonstration E.9 a little push. If the push is vigorous enough, the ball goes into the lower well with a kinetic energy increase. Likewise, if the hemisphere is dropped from a low height, it pops into the other configuration with a gain of kinetic energy, and rises to a much greater height. It would be a good idea not to place your face directly over the point the ball is dropped from, since the hemisphere can jump up with considerable speed. If the hemisphere doesn't jump as expected, either you are dropping it with the wrong side facing down, or it is popping back into its lower-energy state before striking the floor.

This demonstration imperfectly simulates the transition from a high-energy state to a low-energy state in an atom. Following an atomic transition, the liberated energy is in the form of a light photon, while here it is in the form of the kinetic energy of the rubber hemisphere. The hemisphere only begins the transition to the lower-energy state when it hits the ground, so the analogous atomic process would be *stimulated* emission (which occurs in lasers), rather than spontaneous emission. Another demonstration (I.3), involving "jumping disks," simulates the process of spontaneous emission.

# E.11. Path of least time

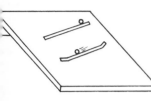

**Demonstration**

The amount of time required for a rolling ball to descend a hill of given height depends on the shape of the hill, even though its speed at the bottom does not, according to energy conservation.

**Equipment**

An 8×10-in. piece of ½-in.-thick acrylic plastic sheet, obtainable at a hardware store or plastics company; and two metal balls.

**Construction**

Cut two pieces of wood or acrylic sheet to form the two ledges on the sheet on which balls are to be rolled (see illustration). Both the top and bottom ledges are ½ × 8 inches, but the top one is straight, and the bottom one has a rise of about 20° at each end. Orient the ledges on the acrylic sheet with one several inches directly below the other, and the right edges about three inches higher than the left edges. A ball rolled down either ledge should experience the same vertical drop from one side to the other. Glue or screw the ledges onto the sheet, and add two screws to the underside of the sheet to prop up the top of the sheet by about an inch when placed on an overhead projector.

**Comment**

Before releasing the balls simultaneously from the upper end of each ledge, you may want to ask for predictions as to which ball reaches the bottom first. People who are familiar with the conservation of energy law, which requires the balls to have equal speeds at the bottom, often mistake that to mean that the times of descent are also equal. But the ball on the lower track reaches the bottom much sooner, since its steep initial descent allows it to cover the bulk of its path at much higher speed than the ball on the upper track. An infinite number of possible paths exist for a ball descending a given vertical distance. One of these paths, known as the brachistochrone, is the path of least time. Clearly, the shape of the lower ledge must more closely resemble the brachistochrone than the shape of the upper one does.

---

# Circular Motion and Angular Momentum

## F.1. Spinning a water-filled cylinder

**Demonstration**
The paraboloidal shape of the water surface in a spinning container is a consequence of Newton's second law.

**Equipment**
A 78-rpm phonograph turntable, and a deep transparent pie dish (or other transparent cylindrical container) at least 7 inches in diameter and 2 to 3 inches deep.

**Comment**
As we will prove, the water surface in a spinning container assumes the shape of a paraboloid described by the equation $y = a(r^2 - b^2)$, where $a$ and $b$ are constants. The paraboloid equation can be used to predict the water rise at the edge, and the corresponding drop in the center, for a cylindrical container spinning at 78 rpm. Applying the equation to four containers of different diameters yields rises of 0.53, 0.69, 0.88, and 1.08 inches at the edge for container diameters of 7, 8, 9, and 10 inches, respectively.

These predictions can easily be checked by putting 1 to 2 inches of water in a deep, transparent pie dish whose diameter is at least 7 inches (the larger the better). The water should be deep enough that the depression in the middle when the dish spins doesn't reach the bottom, but not so deep that the water overflows at the edge. (The preceding numerical predictions should help you choose an appropriate water depth for a container of a given diameter and height.) You may wish to add a bit of food coloring to make the water more visible.

Make three broad horizontal marks on the side surface of the pie dish showing (a) the original water height, (b) the predicted rise of the water level at the water's edge, and (c) the predicted drop of the water level at the center. If magic marker doesn't work well on the glass surface you may want to use transparent tape and make marks on that. When the dish is spun at 78 rpm and viewed from a height level with the water you will find to what extent the prediction is verified. (Remember that if the sides of the pie plate are sloping, you must measure its diameter at the height of the water level.)

In order to derive the paraboloid equation for the shape of the spinning water surface, we first define the direction of "artificial gravity" as the vector sum of the weight of a water droplet, $mg$, and the horizontally acting centrifugal force, $m\omega^2 r$. At any point on the water surface the tangent to the surface must make an angle $A$ with the horizontal that is the same as the angle that artificial gravity makes with the vertical. The tangent of this angle is clearly given by $\tan A = \omega^2 r/g$, so the slope of the water surface is proportional to $r$, the distance to the axis of rotation. Given a paraboloid described by the equation $y = a(r^2 - b)$, we see that $dy/dr = \tan A = 2ar$. This means that the slope of the water surface has the right dependence on $r$ for the surface to be a paraboloid, and, further, that the constant $a$ is given by $a = \omega^2/2g$.

To find the constant $b$ in the paraboloid equation, we require that the volume integral for the solid of revolution defined by $y = a(r^2 - b)$ yield zero volume, because as much water rises up the edges as is taken away from the middle. Thus we let $\int 2\pi ry\, dr = 0$, with $y = a(r^2 - b)$. It follows that $b = aR^2/2$, which gives for the equation of the water surface the paraboloid $y = a(r^2 - R^2/2)$, with $a = \omega^2/2g$.

## F.2. Spinning person on a lazy Susan

### Demonstration
A variety of angular-momentum demonstrations can be given using a homemade rotating turntable.

### Equipment
Two 1-kg masses; a rotating platform; and a weighted bicycle wheel with handles (if available). A very strong and inexpensive rotating platform capable of supporting 300 pounds can be made using a 4-in.-square lazy-Susan ball-bearing plate purchased at your local hardware store. A commercially available "trimmer exerciser" used by people who want to "twist away" excess weight also makes an excellent inexpensive rotating platform.

### Construction
To make the rotating turntable, screw a 14-in.-square board to the underside of the bearing plate and a 14×4×1-in. board to the top side.

**Comment**

A person standing on the rotating turntable will probably have a great deal of difficulty staying on it when given a spin. However, if you assume a low-center-of-gravity crouch with the balls of your feet on the board, you should have little difficulty staying on the platform when someone else gives you a slow spin. If you should get thrown off the spinning platform, your crouching position will ensure that you don't have far to fall, but if you are not an agile person you might want to be the spinner rather than the spinnee. Remember, if you get dizzy, spinning the other way will get you undizzy.

You can illustrate conservation of angular momentum by starting your spin with arms outstretched holding weights, and quickly bringing the weights toward your body. The decrease in your moment of inertia must be compensated for by an increase in your rate of rotation. The constancy of the angular momentum $I\omega$ requires that as $I$ decreases, $\omega$ increases proportionately, so that the kinetic energy $\frac{1}{2}I\omega^2$ must increase (because $\omega^2$ increases faster than $I$ decreases). This increase in your kinetic energy as your moment of inertia decreases is due to the work required to pull the weights toward the axis. You may find surprising the extent to which the weights resist being pulled in as you spin.

A number of other demonstrations of angular momentum conservation can be done using the turntable. For example, if you throw a basketball while standing on the turntable, the angular momentum of the released basketball depends on its velocity and the perpendicular distance between the velocity vector and the turntable axis. Thus, by angular momentum conservation your recoil spin will be much greater if you throw the basketball using a horizontal arm swing than if you shoot it directly forward with both hands.

Several additional demonstrations require a weighted bicycle wheel with handles on the axle. If you hold the rapidly spinning bicycle wheel while standing on the turntable, a change in its axis of spin causes a change in your rotation rate consistent with the vector nature of angular momentum. For example, if the bicycle wheel's axis of spin is initially horizontal, you can make the turntable spin either clockwise or counterclockwise depending on which direction you rotate the bicycle wheel's spin axis to the vertical.

An alternative demonstration starts with the bicycle wheel spinning along a vertical axis while you stand on the stationary turntable. If you hold a handkerchief against the spinning wheel to slow it down, its angular momentum is gradually transferred to you and the turntable. In this case angular momentum is conserved even though rotational kinetic energy is lost.

For a final demonstration using the turntable, consider the problem of a stationary astronaut in space wishing to change her orientation, but having nothing to push against. If she remembers that cats held by their feet are able to land on their feet when dropped, she will realize that it is possible to change her orientation while still conserving angular momentum.

The situation of the astronaut and the cat is similar to yours on the stationary low-friction turntable. To change your orientation, rotate the upper and lower parts of your body in opposite directions and then return to the original position (like doing the "twist"). If you practice a bit you should be able to change your orientation by as much as 45 to 90 degrees each time.

How is this done while conserving angular momentum? Suppose the top portion of your body rotating in one direction—say, clockwise—has a larger moment of inertia than the bottom portion. In this case the top portion of your body has a smaller angular acceleration than the bottom, and it turns through a smaller angle than the bottom in a given time, since equal and opposite torques act on both portions of your body. Before you bring your body back to its original alignment, you must change the average distance of your upper or lower body to the axis of rotation, to change the relative moments of inertia. For example, if you move your arms closer to your body to make the top and bottom parts of your body have more nearly equal moments of inertia, they will rotate through angles that are more nearly equal than originally, with the net result that youhave turned counterclockwise. As when doing the "twist," relying on your body's intuition may produce the best results.

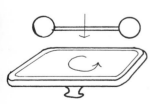

## F.3. Dropping objects onto a rotating casserole cover

### Demonstration

The remarkably low friction of an inverted spinning casserole cover allows you to test angular momentum conservation by dropping objects onto the cover.

### Equipment

A glass casserole cover—in particular, one that has a handle with a hollowed-out depression in the center; a pair of large clay balls; and a straw. Put the clay balls on the ends of the straw, making a dumbbell whose length is a bit less than that of the casserole cover's diameter. The dumbbell's weight should not be small compared to that of the casserole cover, and the clay balls should be flattened so that they won't roll. Put a piece of masking tape along a radius of the cover to make the period of rotation more easily observable when the cover is spun on an overhead projector.

**Comment**

When the casserole cover is inverted and spun, its remarkably persistent rotation is a consequence of the rocking motion that arises due to the shape of the cover's handle. The rocking results in momentary surface contact of ever-changing points on the handle, as in rolling—rather than sliding—friction. You need to be careful when you spin the casserole cover, because it has a tendency to "walk," so stand ready to catch it if it walks off the edge of the surface.

If you spin the casserole cover and drop the dumbbell from a low height squarely onto the cover, you will observe an appreciable decrease in spin rate. But if you repeat the demonstration with the two clay balls next to the center of the straw, you will observe a much smaller change in spin rate. Conservation of angular momentum requires that a change in moment of inertia result in a compensating change in spin rate. Dropping the dumbbell onto the spinning cover changes the moment of inertia more when the clay balls are at the ends of the straw than when the balls are near the center of the straw, so there is also a greater change in angular speed.

## F.4. Tornado in a bottle

**Demonstration**

Swirling a water-filled inverted jug that has a hole in its cap causes a tornado-like whirlpool to form in the jug, which surprisingly persists until the jug has drained, because the water in the jug extracts angular momentum from the water that leaves.

**Equipment**

A gallon jug with a ⅜-in. hole drilled in its cap, and a tray to catch a gallon of water.

**Comment**

Fill the jug about ⅔ full of water and turn it upside down. Rapidly swirl the liquid around in the jug, which will cause a small whirlpool to form. If you have swirled it hard enough, a "tornado" will form in the jug, and it will last the whole time the jug is draining—perhaps over a minute. If you have difficulty getting the tornado to form, try enlarging the hole in the cap.

The persistence of the tornado may be puzzling, because when water is swirled in a jug that is not draining, the rotation of the water stops shortly after you swirl it. The difference is that the water that leaves the draining jug gives up some of its angular momentum to the remaining water, thereby counteracting the loss of angular momentum due to friction. As the swirling water approaches the narrow mouth of the inverted jug (and lowers its moment of inertia), its rotation rate would need to speed up appreciably for its angular momentum to stay constant. This cannot occur, since it would require a spontaneous increase in the descending water's kinetic energy in excess of its loss of gravitational potential energy. (Recall that pulling in a pair of weights while spinning on a rotating platform requires work.) Therefore, the water can exit the narrow mouth of the bottle only by giving up some of its angular momentum to the water above, causing the remaining water to persist in its swirling motion. The demonstration also works without using a bottle cap, but then the jug drains too quickly.

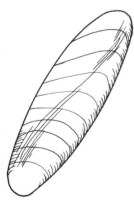

## F.5. "Rattleback"

**Demonstration**
Angular momentum conservation appears to be violated by an oddly shaped stick that reverses its direction of spin only when spun in one direction, an effect that is due to coupled oscillations.

**Equipment**
A piece of wood carved to have a cross section that varies as shown (or purchased from Jerryco or Toltoy Inc, under the name "Space pets," for about $1.00).

**Comment**
Giving the rattleback a rocking motion will cause it to begin to rotate in one direction, because of the peculiar shape of its bottom. Conversely, a rotation in this direction contributes to a rocking motion, but in a cumulative manner only when the periods of rocking and rotation match. This two-way coupling between rocking and rotation is the rea-

son the rattleback begins to rock once its spin rate slows sufficiently, and the reason the rocking in turn gives rise to a small spin in the opposite direction. No such peculiar behavior occurs if the rattleback is spun in the opposite direction, since the asymmetrical shape of its bottom causes the rocking motion to be coupled to a rotation in only one direction.

## F.6. "Whirligig" with a hanging weight

### Demonstration
A twirling weight, connected to a hanging weight through a tube, can be used to show angular momentum conservation, and to verify the relation $a = v^2/r$.

### Equipment
A tube; two weights connected by a string; and a metronome or a clock.

### Construction
The proper construction of the "whirligig" is trickier than you might think! First, the tube should be made of a solid plastic or wooden dowel, in which a ¼-in.-diameter hole is drilled for the string. The dowel should be long enough that you can hold it in your hand and twirl the string, but short enough that you can drill through from one side only. In order to avoid friction when the string rubs against the top of the tube, you will need to bevel both the edge of the hole and also the outer edge of the dowel. Plastic is better than wood, since it is smoother, but if you use wood, sand the surface on which the string will rub, and make sure it is as smooth as possible.

Second, the choice of the two weights is important. You should use a light ball for the twirled weight and a ball about three times as heavy for the hanging weight. The twirled ball should be light for the following reasons: (a) if the string breaks, only a light object goes flying off; (b) the lighter the spinning ball, the closer the angle $A$ between the string and the vertical approaches 90°, and hence there is less friction between the string and the tube; and (c) the mathematical analysis is easier if $A$ is close to 90°. However, the hanging ball should not be much more than three times the weight of the other ball; otherwise, for the hanging weight to supply the required centripetal force, too high a rotational speed is required.

**Circular Motion &
Angular Momentum**

**Comment**

You can demonstrate angular momentum conservation by simply pulling down on the hanging string while the light ball twirls in a horizontal circle. As the light ball is moved toward the center, its moment of inertia decreases and its rate of rotation will increase significantly. In order to verify the relation $a = v^2/r$, you can keep the acceleration $a$ constant, and show that $v^2$ is proportional to $r$. This can be done most easily for the case in which angle $A$ is near 90°, since the hanging weight (if it is stationary) provides a constant centripetal force, and therefore a constant centripetal acceleration.

The observation you need to make is as follows: Rotate the light ball at a fixed period—say, 1 second—either by using a metronome synchronized to the rotation or by having someone count off the seconds while you keep the rotation rate constant. You also need to see how much string hangs down below the bottom of the dowel, so you can use this length afterward to determine the radius of the circle for the rotating ball. Be sure to wait until the hanging weight assumes its equilibrium height for the particular rotation rate.

With the measured period $T$ and radius $r$ you can calculate the ball's velocity from $v = 2\pi r/T$ and its acceleration from $a = v^2/r$. The measured acceleration can then be compared with that calculated from the equation $a = Mg/m$, which is the force that in the absence of friction acts on the ball (the value of the hanging weight, $Mg$) divided by the ball's mass $m$. However, because of friction there may be a significant discrepancy between the two values for the acceleration.

An alternative comparison, which also verifies $a = v^2/r$ and for which friction is of little importance, is to look at several different rotation rates and the associated radii. Using a metronome, twirl the ball at a number of different rates, and find the velocity and radius in each case. You should discover that a plot of $v^2$ versus $r$ yields a straight line, thereby verifying the relation $a = v^2/r$ for the case of constant acceleration.

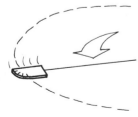

## F.7. Twirling an accelerometer in a nearly horizontal circle

**Demonstration**

By twirling an accelerometer in a nearly horizontal circle, you can verify the relation $a = v^2/r$.

74

**Equipment**

A clock, and an "impact-stress meter" sold for about $20.00 (see demonstration A.5). This device can be set to beep when the acceleration is any value between 2.3 and 5.0 $g$'s.

**Comment**

Set the accelerometer dial to 5 $g$'s, and twirl it at the end of a string in a nearly horizontal circle. Keep the rotation period fixed at one second by matching your rotations to the seconds counted off by someone. Vary the length of the string until the device is just on the verge of beeping (any shorter, it won't beep at all; any longer, it beeps continuously). As shown below, you should find that the string length for which the device is on the verge of beeping is 1.24 meters minus half the length of the device.

This result follows if we let the acceleration $a = v^2/r = 5g$, and substitute $v = 2\pi r/T$. Solving for $r$ yields $r = 5T^2g/4\pi^2$ = 1.24 meters, assuming $T = 1$ second. An alternative to verifying the 1.24-meter radius would be to use the measured radius to compute the acceleration from $v^2/r$, and see how close you come to the value $5g = 49$ m/s$^2$.

The use of the maximum acceleration $a = 5g$ ensures that the angle with the vertical, $A$, is close to 90° (actually 78.3°). For smaller accelerations and angles you need to correct for the fact that the length of the string and the radius are not the same, but are related by $r = L/\sin A$ (which for $A = 78.3$° is only a 2 percent correction).

## F.8 Conical pendulum

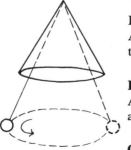

**Demonstration**

A conical pendulum makes an angle with the vertical related to its length and period through Newton's second law.

**Equipment**

A pendulum on a string; and two cardboard cones, one with a 30° half-angle and the other a 48.6° half-angle.

**Construction**

To make the two cones, cut a cardboard file folder into two 8-in.-diameter disks. Cut the first disk in half along a diameter, and roll up one semicircular piece into a cone; its half-angle will be 30°. Seal the seam of the cone with tape. Cut a quarter of a circle out of the remaining disk. When you roll up the larger piece to make a cone, its half-angle will be 48.6°. Make a small hole in the apex of each cone through which to pass the pendulum string.

**Comment**

We can easily prove that only conical pendulums of a certain length can make a particular angle with the vertical if their period is also fixed. For example, we later show that conical pendulums with a one-second period must have lengths of 0.286 meters and 0.184 meters, if they are to have half-angles of 30° and 48.6°, respectively.

These results can easily be checked by experiment. Place the pendulum string through the hole in the top of the 30° cone, and gently twirl the pendulum, holding the top of the cone and string with two fingers and keeping the cone axis vertical. You may find it helpful to put a small piece of clay at the inside apex of the cone to prevent distortion of the cone's shape by your finger pressure.

Synchronize your rotation rate to the sound of someone counting out the seconds, and vary the length of the string until the pendulum makes the same angle with the vertical as the sides of the cone. The string length is just right when any shorter length causes the pendulum to lie inside the cone, and any longer length causes the string to rub on the edge of the cone. Compare your measured string length for the 30° cone with the above prediction, and repeat the procedure using the 48.6° cone.

To derive the cited results for the two pendulum lengths, we begin by noting that the resultant of the forces on the pendulum bob, obtained by adding the weight and string tension, must point toward the center of the circle in which the bob moves, and must have a magnitude $m\omega^2 r$. This is indeed the case provided the string's angle with the vertical satisfies the relation $\tan A = \omega^2 r/g$, where $\omega$, the angular speed, is related to the period $T$ through the equation $\omega = 2\pi/T$. Solving the preceding equation for the radius $r$ yields $r = (gT^2\tan A)/4\pi^2$. In the case of two pendulums with one-second periods, and with half-angles $A = 30°$ and $A = 48.6°$, we obtain the radii $r_1 = 0.143$ m and $r_2 = 0.0641$ m, respectively. The pendulum lengths corresponding to these radii are $L_1 = 0.286$ m (for 30°) and $L_2 = 0.184$ m (for 48.6°).

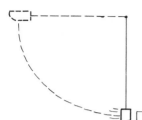

# F.9. Accelerometer pendulum

**Demonstration**
When swung through a 90° angle, an accelerometer is observed to experience 3 $g$'s at the bottom of the swing, according to $F = mv^2/r$.

**Equipment**
Same as demonstration F.7.

**Comment**
For a pendulum that starts at rest at a 90° angle to the vertical, energy conservation requires that the speed of the pendulum at the bottom of the swing be consistent with the relation $\frac{1}{2}mv^2 = mgr$, or $v^2 = 2gr$. The force the accelerometer experiences at the bottom of the swing is given by $F = mg + mv^2/r$, which works out to be $3mg$ (or 3 $g$'s). You can measure the acceleration at the bottom of the swing by determining the exact accelerometer setting at which the accelerometer first beeps, and compare the result with 3 $g$'s.

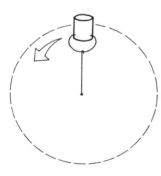

# F.10. Whirling a water-filled can in a vertical circle

**Demonstration**
The minimum rotational period for a water-filled can whirling in a vertical circle—without a wet mess being created—can be predicted from the equation $F = mv^2/r$ and from energy conservation.

**Equipment**
A "tin" can.

**Construction**
Drill three holes in the can near the top through which to tie a string on.

**Comment**
At the top of the circle, if the speed of the can is sufficiently high, both the water and the can have a *common* downward acceleration in excess of $g$, so the water doesn't fall out. The acceleration of the can decreases if you shorten the string while keeping the rotation period fixed. As shown below,

the minimum length of string that would allow you to whirl a can in a vertical circle with a 1-second period of rotation is 0.63 meters.

You can verify this result by whirling a can in a vertical circle while synchronizing your rotations to the sound of someone calling out the seconds. Start with a one-meter-long string, and gradually shorten it. You should find that as you approach the point at which the length of the string plus half the height of the can is 0.63 meters, the water is on the verge of falling out of the can at the top of the circle.

In order to derive the $r = 0.63$ m result, we note that the centripetal acceleration of the can at the top of the circle must be no less than $a = v^2/r = g$ or the water will fall out, from which we find $v_{top} = (rg)^{1/2}$. At the bottom of the circle, the can increases its kinetic energy by $mgh = mg(2r)$, so its speed is greater, and is given by $v_{bot} = (5rg)^{1/2} = 2.24(rg)^{1/2}$. The average speed of the can around the circle approximately equals the average of its speed at the top and bottom, or $v_{avg} = 1.62(rg)^{1/2} = 5.08r^{1/2}$ m/s, using $g = 9.8$ m/s².

The time for one revolution can be expressed as $T = 2\pi r/v_{avg} = 1.23r^{1/2}$ seconds, where $r$ is in meters. Solving this equation for $r$ yields $r = 0.66T^2$, which for $T = 1$ second gives $r = 0.66$ meters. Had we not made the approximation for the average speed around the circle and instead calculated the time for one revolution exactly by integration, the answer is only slightly less: $r = 0.63$ meters.

## F.11. Gyroscope

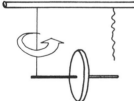

**Demonstration**
The precession of a spinning gyroscope when one of two strings supporting it is cut shows the vector nature of angular momentum, and the relationship between torque and rate of change of angular momentum.

**Equipment**
A gyroscope, available from Toys-R-Us for about $5.00.

**Construction**
Suspend the gyroscope from two strings hanging from a horizontal bar. Be sure that the two vertical supports for the bar are far enough from the strings that the gyroscope doesn't strike the supports when precessing. You may want to put this arrangement on top of an overhead projector if giving the demonstration to a group. A weighted bicycle wheel

with handles can also be used in a large group setting in lieu of the gyroscope.

**Comment**

First show that the non-spinning gyroscope falls when one of the strings is released. Then give the gyroscope a hard spin, and cut one of the two supporting strings. The gyroscope will precess about the other string. That the gyroscope does not fall is a wonder to behold, and can be explained most easily using the vector nature of angular momentum, $\vec{L}$.

For example, a gyroscope spinning as shown in the illustration has an angular momentum vector pointing to the right (along the rotation axis), according to the right-hand rule. This is shown in Plate F.11a. Now, if the right-hand string is cut, gravity exerts a torque $\vec{\tau}$ on the gyroscope, given by $\vec{r} \times \vec{F}$, which points *into* the plane of the paper. According to Newton's second law, the change in the gyroscope's angular momentum $d\vec{L}$ must also point into the plane of the paper. As seen in Plate F.11a, the initial angular momentum vector $\vec{L}_o$ will rotate about the vertical axis in a counterclockwise direction.

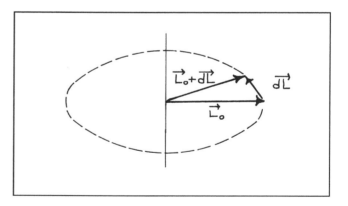

**F.11a** Angular momentum vectors for a precessing gyroscope

## F.12. Pennies on a rotating turntable

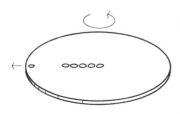

**Demonstration**

Whether a penny stays on or flies off a rotating turntable can be determined from the relation $F = mv^2/r$.

**Equipment**

A stopwatch; a large rotating turntable (the kind used as a kitchen organizer); and a number of pennies.

**Comment**

Static friction provides the centripetal force on the pennies. For a row of pennies along a radius of the turntable, the outermost one in the row that doesn't fly off when the turntable is spun experiences the maximum force of static friction, $\mu mg$, which also equals $mv^2/r$. If we observe, for various rotation rates, which penny is the first not to fly off, we can see how well the penny's distance to the axis, $r$, and its velocity $v$ satisfy the relation $v^2/r = \mu g$ (a constant).

To conduct the demonstration, first tape a piece of paper on the turntable and locate the turntable's exact center by seeing where a penny can be placed and not appear to wobble as the turntable is rotated. Draw a circle around the penny, and draw a row of similar, adjacent circles around pennies placed along a radius. Number each circle and measure the distance $r$ from the center of each penny to the rotation axis.

Rotate the turntable at some speed, and see which pennies fly off. Also, record the time for one revolution, $T_0$. Since the rotation rate will slow after a few turns, you want to measure the time for the first one or two revolutions.

Suppose that $r_0$ is the distance to the axis from the center of the outermost penny that does not fly off. You can compute its velocity from $v = 2\pi r_0/T_0$, and its centripetal acceleration from $a = v^2/r_0$.

If you repeat this demonstration for a different rotation rate and measure the new rotation period $T_1$, you can see if the acceleration of the outermost penny is the same as that computed previously. The same acceleration should be found regardless of rotation rate, because the static friction force $f = \mu mg$ is the centripetal force agent, and so the acceleration is a constant: $a = f/m = \mu g$ for all pennies that are just on the verge of slipping.

One related demonstration you can give with the rotating turntable would be to replace the pennies with a number of cylinders of different height-to-diameter ratios, and see which of them tip over—rather than slide—when the turntable is spun. A cylinder should tip rather than slide if the height-to-diameter ratio $h/d$ exceeds the reciprocal of the coefficient of static friction. This result, which is independent of rotation rate or the cylinder's distance from the axis, can be obtained by comparing the conditions for sliding [$mv^2/r = \mu mg$] and tipping [$mv^2/r = (d/h)mg$]. In order to carry out this demonstration, you need to prepare a number of cylinders whose height-to-diameter ratios bracket the reciprocal of the coefficient of static friction, using the method described in demonstration D.5.

# F.13. Rolling a ball on a rotating turntable

### Demonstration
Rolling a steel ball straight across a piece of carbon paper on a rotating turntable yields a curved path that can be explained as the result of two "fictitious" forces, the Coriolis force and the centrifugal force.

### Equipment
A large rotating turntable; a steel ball; and carbon paper (pencil carbon paper is the most sensitive type). If you want to give this demonstration to a large group, you will also want a blank transparency. In that case, be sure to use a heavy steel ball to get good, dark trails. A ball diameter of 1.5 inches would be a good size.

### Comment
Most people are more familiar with the centrifugal force than with the Coriolis force, but both can occur in a rotating frame of reference. We can investigate both forces using a rotating turntable on which carbon paper is placed, face up, underneath a piece of ordinary paper (or a transparency blank, if you want to use an overhead projector to show the results). Locate the center of the turntable and make a mark there. Spin the turntable and roll the ball across the paper. Notice that the ball—from your perspective—rolls approximately in a straight line. (The line is not exactly straight, however, because at high rotation rates the ball acquires some sideways spin, which causes a deflection.)

In the frame of reference of the rotating turntable (what the carbon paper records), the ball's path differs dramatically from a straight-line path, as you can easily see by examining the carbon-paper trails for various rotation rates and ball velocities. In the laboratory frame of reference the ball experiences zero net force apart from friction, but in the rotating reference frame the ball experiences two "pseudoforces"—the Coriolis force and the centrifugal force. The centrifugal force always points outward along a radius vector; the Coriolis force acts at right angles to the ball's velocity and deflects the ball to one side.

By examining the carbon-paper trails for balls rolled on both sides of the center point, it is easy to see which of the two pseudoforces is dominant. If the centrifugal force is dominant, the paths will always curve away from the center regardless of the side the ball is rolled on. If the Coriolis force is dominant, the balls should curve toward the center when rolled on one side, and away from the center on the other side (see Plates F.13a and F.13b). As a general rule,

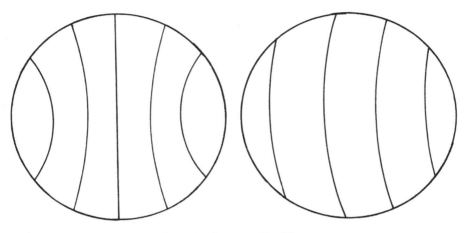

**F.13a** (left)  Path of a rolling ball acted on by a centrifugal force
**F.13b** (right)  Path of a rolling ball acted on by a Coriolis force

you can expect the centrifugal force to dominate when the turntable rotation time is much less than the time the ball takes to roll across the turntable, and the Coriolis force to dominate when the opposite is true.

## F.14. Transfer of angular momentum

### Demonstration
According to law of conservation of angular momentum, the loss in angular momentum of water swirling in a gallon jug results in a gain in angular momentum of the rotating platform on which the jug is placed.

### Equipment
A clear gallon jug, and a rotating turntable. (A two-liter soda bottle cannot be used in place of the jug, because it doesn't have a large enough diameter.) An alternative version of the demonstration using a raw egg is suitable for an overhead projector.

**Comment**

Fill the jug about two-thirds full of water and swirl the water around very vigorously. Before the water has a chance to lose much angular momentum, quickly place the jug at the center of the stationary turntable. As the water slows down, the turntable and jug will begin to spin in the same direction the water was rotating, and then soon come to rest because of friction. A very vigorous swirling is necessary to get a readily observable effect.

A simpler version of this demonstration requires only a raw egg. Give the egg a vigorous spin. Momentarily stop the egg, and then let it go. It should begin spinning on its own, because the contents of the egg retain their angular momentum even when the egg itself has momentarily been brought to rest. As the contents lose their angular momentum due to viscous forces, angular momentum conservation requires that the egg itself begin to spin.

## F.15. Pennies on a falling meterstick

**Demonstration**

When a meterstick with a row of pennies on it falls, while remaining supported at the 0-cm end, only the pennies up to the 66⅔-cm mark remain in contact with the stick, according to Newton's second law for rotation.

**Equipment**

A meterstick, and 20 or more pennies.

**Comment**

The easiest way to show the demonstration is to support one end of the stick on the edge of a desk and suddenly let the other end go. You may want to put a quarter at the 66⅔-cm point to indicate the place at which the coins should leave the stick. The pennies on the last third of the meterstick lose contact because this part of the stick has a downward acceleration greater than $g$.

We obtain this result if we divide the torque about one end, $mgL/2$, by the moment of inertia, $\frac{1}{3}mL^2$, to obtain $1.5g/L$ for the angular acceleration. The linear acceleration must therefore exceed $g$, provided the distance $r$ to the pivot exceeds $\frac{2}{3}L$.

If you find it too tedious to line up the pennies on the stick and to pick them up from the floor afterward, a simple way of doing the demonstration is to place a penny or a ball at various locations on the stick before releasing one end.

## F.16. Angular rotation of a beam

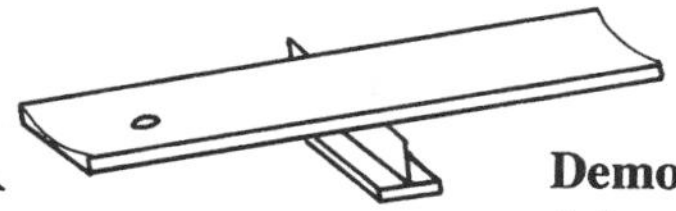

**Demonstration**
A long, balanced beam swings with an angular acceleration that is proportional to the applied torque, according to Newton's second law for rotation.

**Equipment**
A beam consisting of a 6-foot-long, $1\frac{5}{8}$-in.-wide piece of wooden molding with a cross section like that shown in the figure; a knife-edge on which to balance the beam; pennies; and a stopwatch.

**Construction**
Rest the center of the beam on the knife-edge (concave side up) so that it balances as precisely as possible. Press the beam onto the knife-edge to score it at the balance point; this will make it easier to reposition the balanced beam. The knife-edge should be placed on a 6-in.-high support, so that the ends of the beam can swing 6 inches before hitting the table. Make a 6-in.-high wooden block to place under one end of the beam, as a means of seeing that the beam is initially horizontal when it starts its swing. Make marks along the top of the beam spaced every 6 inches from a center line.

**Comment**
A simple procedure exists for measuring the angular acceleration of the balanced beam when it is acted on by a torque. Place the 6-in. block under one end of the beam so that the beam is horizontal when its end rests on the block. Add a penny to the top of the beam, placing it on one of the lines on the side of the beam that is resting on the block. Move the block out from under the beam without disturbing it, and time the beam's swing to the table from the instant you withdraw the block.

For a swing through a fixed angle $A$, the square of the time of the swing is inversely proportional to the angular acceleration, based on the equation $A = \frac{1}{2}\alpha t^2$. The torque acting on the beam due to the weight of the penny is directly proportional to the penny's distance from the center. Therefore, according to Newton's second law for rotation, an inverse proportionality should exist between the penny's distance from the center $r$ and the square of the swing time $t$.

You can easily test this relationship by timing the beam's swing when the penny is placed at each of the lines, on both sides of the balance point. Average five swing times for each $r$ value to reduce random error. Newton's second law for rotation will be confirmed if you find that a plot of $r$ versus $1/t^2$ is a straight line through the origin.

## F.17. Ball rolling on a balanced beam

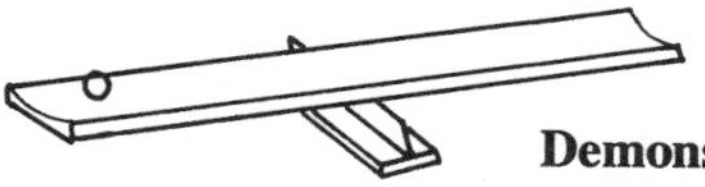

### Demonstration
A Ping-Pong ball rolling in a groove on a balanced beam exhibits oscillatory motion only for precisely determined initial conditions.

### Equipment
A 6-foot-long piece of wooden molding (see demonstration F.16); a knife-edge on which to balance it; and a Ping-Pong ball. A ball can be rolled on the molding because of the cross-sectional shape.

### Comment
The weight of a ball placed in a groove on a balanced beam causes the beam to swing, and the beam's angle, in turn, affects the acceleration of the ball. Thus the ball and the beam are a coupled system.

As shown below, for any initial angle $A$ of the beam, there exists only one initial ball position—a "critical point"— for which the ball and the beam oscillate indefinitely. We will show that for small angles the critical position of the ball, measured from the center of the beam, is given by $x_0 = 0.189LA_0(M/m)^{1/2}$, where $L$ is the beam's length, $A_0$ is the initial angle of the beam in radians, $M$ is the mass of the beam, and $m$ is the mass of the ball.

If the ball is released from a point other than the critical point, it rolls off one end of the beam after some number (possibly zero) of oscillations. Clearly the closer the ball is placed to the critical point, the more oscillations it makes before rolling off one end. In practice, however, it is not possible to achieve many oscillations before the ball rolls off, owing to the extreme sensitivity of the motion to the initial conditions. Achieving one oscillation is easy, achieving two is difficult, and achieving three is extremely difficult.

To test the equation for $x_0$, first tilt the balanced beam so that one end of the beam rests on the table, and the other end is a distance $h$ above the table. $A_0$ in radians is approximately given by $h/L$. Calculate $x_0$, and place the ball on the raised side of the beam a distance $x_0$ from the center. See how many oscillations occur before the ball rolls off one end. By trial and error, make small adjustments in the initial location of the ball to obtain the largest number of oscillations. By trial and error I found that the initial position of the ball that gave the largest number of oscillations (three) was within an inch of the computed location—reasonably good agreement, considering the uncertainties in the beam's initial angle and in the mass of the ball.

In order to show that only one initial position of the ball leads to oscillatory motion, we need to find the beam's equation of motion. As shown below, the general solution for the angle of the beam as a function of time involves four functions: $\sin \omega t$, $\cos \omega t$, $e^{+\omega t}$, and $e^{-\omega t}$. The general solution is a linear combination of these four functions, so it involves four constants determined by four initial conditions: the ball's initial position and speed, and the beam's initial orientation and angular speed. For an arbitrary choice of initial conditions, the exponentially growing function $e^{+\omega t}$ will contribute to the solution with a nonzero coefficient. In this case, the solution will diverge, and any initial oscillatory behavior will be short lived.

A consequence of the presence of some component of the growing exponential function is that the ball and the beam exhibit chaotic behavior, just like the pendulum of demonstration C.14. For example, if the ball has an initial position that is infinitesimally close to the critical point, it is not possible to predict how many oscillations the ball will undergo before it rolls off, unless the size and direction of the infinitesimal distance are specified—which is not possible in an actual physical situation. The remainder of this discussion is a derivation of the beam's equation of motion, and the solution of this equation. This derivation, which some readers may wish to skip, involves a fourth-order differential equation.

According to Newton's second law for rotation, the angular acceleration of the beam is given by $\alpha = \tau/I$, where $\tau = mgx \cos A$, and $I = ML^2/12$. Thus we have $\alpha = D^2A = (12mgx \cos A)/ML^2$, where we use the notation $D^nA$ for the $n$th derivative with respect to time. For small angles, we assume $\cos A \approx 1$, so that we obtain

$$D^2A = 12mgx/ML^2. \tag{1}$$

Differentiating equation (1) twice yields

$$D^4A = (12mg/ML^2)D^2x. \tag{2}$$

The acceleration of a *hollow* ball on an incline of angle $A$ is

$$a = D^2x = \tfrac{3}{7}g \sin A \approx \tfrac{3}{7}gA \tag{3}$$

for small $A$. Substituting equation (3) into (2) yields

$$D^4A = (5.14mg^2/ML^2)A, \tag{4}$$

which can be simplified to

$$D^4A = \omega^4A, \text{ with } \omega = (5.14mg^2/ML^2)^{\frac{1}{4}}. \tag{5}$$

As can be shown by direct substitution, the four particular solutions of equation (5) are $\sin \omega t$, $\cos \omega t$, $e^{+\omega t}$, and $e^{-\omega t}$. If we want the solution to be periodic with a maximum angle at $t = 0$, we let $A = A_0\cos \omega t$, so that we have

$$D^2A = -\omega^2A_0\cos \omega t. \tag{6}$$

Combining equations (1) and (6), and letting $t = 0$, we have

$$12mgx_0/ML = -\omega^2A_0. \tag{7}$$

Solving equation (7) for $x_0$ gives the ball's initial position:

$$x_0 = -\omega^2A_0ML^2/12mg. \tag{8}$$

Using the value of $\omega$ from eqn. (5), we can rewrite eqn. (8) as

$$x_0 = -0.189LA_0(M/m)^{\frac{1}{2}}, \tag{9}$$

thereby proving the result stated earlier. Had the derivation for $x_0$ been done for a solid sphere rather than for a hollow one, the only difference would have been that the constant in equation (9) would have been −0.244. The minus sign in equation (9) is only significant in that it gives the relative signs of $x_0$ and $A_0$.

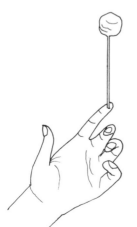

## F.18. Balancing a stick on its end

### Demonstration
Newton's second law for rotation explains why long sticks are easier to balance than short ones, and why a weight on the top end of a stick makes balancing a stick easier.

### Equipment
A 12-inch ruler; a meterstick; a pencil; and some clay.

### Comment
Your ability to balance a stick on your finger depends on the stick's length, as you can easily verify by trying sticks of different lengths. You should find it easy to balance a meter-

stick, hard to balance a 12-inch ruler, and impossible to balance a pencil. A directly related demonstration is that long sticks placed on their end on a table take longer to topple over than short ones.

Both demonstrations can be explained by Newton's second law for rotational motion, $\tau = I\alpha$. A uniform stick that makes an angle $A$ with the vertical experiences a torque due to gravity about its bottom given by $\frac{1}{2}mgL \sin A$. The moment of inertia about the bottom is $\frac{1}{3}mL^2$, so the stick's angular acceleration is $(3g/2L)\sin A$. Long sticks, therefore, have a smaller angular acceleration, and they are easier to balance than short ones.

Suppose we now consider a nonuniform stick—for example, a stick with a clay ball on one end. If the clay ball is on the top end, the stick is easier to balance, while if it is on the bottom end the stick is harder to balance. This can be verified by trying to balance the 12-inch ruler with a clay ball on its end. Since it is hard to balance without the clay ball, adding the ball to the top makes balancing easier, while adding it to the bottom makes it impossible. Similarly, given two rulers, one of which has a clay ball on its end, the ruler with the clay ball topples over more slowly if the clay-ball end is up, and it topples over faster than the other ruler if the clay-ball end is down.

These observations can also be explained by Newton's second law for rotational motion. Adding the clay to the top of the stick increases both the torque on the stick due to gravity and the moment of inertia. It can easily be shown, however, that the fractional increase in moment of inertia is greater than the fractional increase in the torque, so the value of the stick's angular acceleration is lower with the clay added. For example, it can easily be proved that the angular acceleration of a light stick with a heavy clay ball on its top end is 33 percent less than the angular acceleration of the stick with no clay ball. Conversely, if the clay is added to the bottom of the stick, the fractional increase in moment of inertia is less than the fractional increase in the torque, and the stick's angular acceleration is greater.

## F.19. Moment of inertia of a hula hoop

### Demonstration
A weighted hula hoop can be given an angular acceleration much more easily when an axis of rotation is chosen that is close to the weights.

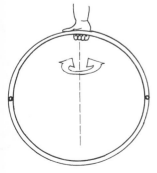

**Equipment**

A hula hoop, inside which steel or lead weights have been inserted at points 180° apart. For best results cut the hula hoop into halves; the halves can be rejoined using steel rods that fit snugly inside each section. An alternative version, suitable for use on an overhead projector, can be constructed using two 3×5 index cards and some pennies.

**Comment**

The moment of inertia of the hula hoop, about an axis in the plane of the hoop that passes through the hoop's center, is smallest when the axis passes through the two weights, and greatest when the axis is perpendicular to the line joining the two weights. Consequently, if you hand the hoop to someone and ask him to twist it back and forth in one hand as fast as he can, he will achieve far more oscillations per second if he grasps the hoop at one of the weights than if he grasps it at a point midway between the weights. The effect is quite puzzling if you haven't mentioned anything about the weights.

An alternative version of this demonstration, which dispenses with the hula hoop and is suitable for use on an overhead projector, uses four pennies and two 3×5-in. index cards. Cut each card into the shape of a "T," making the vertical part 4 inches long and ⅛ inch wide, and the horizontal part 3 inches long and 1 inch wide. Tape two pennies near the center of the horizontal part of one T, and tape two pennies near the right and left edges of the horizontal part of the other T. The center-to-center distance between pennies for the second T should be twice that for the first one. If you grasp the bottom of one T with two fingers and let it hang inverted above an overhead projector, you can easily display its torsional oscillations by twisting it back and forth in resonance with the natural frequency. The T with the pennies far from the axis of oscillation has four times the moment of inertia of the other one and will oscillate with twice the period, because the period of torsional oscillations is proportional to $I^{1/2}$.

# Mechanical Oscillations and Resonance

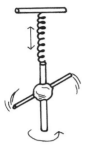

### G.1. Wilburforce pendulum

**Demonstration**

The coupling between linear and torsional oscillations can be demonstrated using a Wilburforce pendulum.

**Equipment**

A spring with a small force constant (a number 46 spring should be suitable); a 1-foot-long, ⅜-inch-diameter steel rod; a ball of clay; and a straw.

**Construction**

Be sure that the end of the spring fits *snugly* into a small hole drilled into the ⅜-inch rod near the top. The device will not work if the spring is merely tied onto the rod. Also be sure the weight is heavy enough that the spring coils don't collide when the mass moves upward.

**Comment**

You want to find a mass/spring combination for which the periods of linear and torsional oscillations are equal. In general, the ratio of the two periods depends on the radius of gyration of the suspended mass. The proper radius of gyration is achieved by starting with the hanging ¾-inch steel rod (whose radius of gyration is too small, since torsional oscillations are faster than linear) and adding a large lump of clay at the center of the rod to increase the radius of gyration. By molding the clay, you can change the radius of gyration as necessary. A straw through the lump of clay makes the torsional oscillations easier to observe.

When the system is far from resonance, measure the amount of time necessary for 50 oscillations of each type in order to see whether to decrease the radius of gyration (if the period of torsional oscillation exceeds the period of linear oscillation) or increase it (if the reverse is true). Near resonance the torsional and the linear oscillations will simultaneously become excited. An interesting phenomenon occurs when there is a slight difference in the frequencies. If you excite the torsional oscillations by twisting the mass and then letting go, you will find that the torsional oscillations gradually diminish, and the energy is fed into the linear os-cillation mode. As time goes by, the energy feeds back and forth between the two modes. The two periods must be fairly close (within about 2 percent) in order for you to observe this most clearly, as explained in the discussion of coupled pendulums (demonstration G.7). You may find that a third oscillation mode, namely simple pendulum oscillations, is also excited, depending on how close its frequency is to the frequencies of the other two modes.

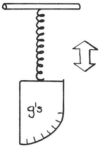

## G.2. Accelerometer in simple harmonic motion

**Demonstration**
An accelerometer oscillating in simple harmonic motion can measure the maximum acceleration that occurs at the end points of the motion.

**Equipment**
A homemade Ping-Pong ball accelerometer (described in demonstration C.9); or a commercially available "impact-stress meter" (see demonstration A.5) and a spring with suitably small force constant. (Actually, a rubber band may be better because the accelerometer is very light and may cause the spring coils to collide during the upward half of the oscillation.)

**Comment**
When the "stress meter" accelerometer undergoes simple harmonic motion on the end of a rubber band or spring, its acceleration can be expressed as a function of its displacement from equilibrium, $x$, and the period $T$ according to $a = 4\pi^2 x/T^2$. Obviously, the acceleration is a maximum at the extremes of the oscillation, where $x$ equals $\pm x_0$, the oscillation amplitude.

You should set the accelerometer to beep at some particular acceleration and by hand vary the oscillation amplitude until you just hear the beep at the bottom of the oscillation. Measure the oscillation amplitude $x_0$ (half of the total vertical displacement) that just makes the device beep, and also record the oscillation period $T$ (the time for one *full* oscillation). You can compare the acceleration calculated from $4\pi^2 x_0/T^2$ to the value on the accelerometer dial, but re-member that gravity adds one extra "$g$" for vertical oscillations (which is why you hear a beep at the bottom of each oscillation but not the top). If you have trouble getting the accelerometer to beep for any oscillation amplitude, try reducing the number of $g$'s set on the dial.

If you use the Ping-Pong-ball accelerometer instead, you can move it by hand in approximate simple harmonic motion by sliding it back and forth on a table, synchronizing its period to the sound of someone counting off the seconds. You can best approximate simple harmonic motion by visualizing your forearm as a pendulum swinging back and forth; try not to move your forearm abruptly. By gradually increasing the oscillation amplitude while keeping the period fixed at one second, you can hear at what amplitude the Ping-Pong ball first pings the side of the jar at the extremes of the oscillation.

The ball strikes the jar when it has an acceleration $a = g \tan A$, where $A$ is the angle through which the ball is free to swing. The acceleration occurring during simple harmonic motion is $a = 4\pi^2 x_0/T^2$, so therefore the amplitude must be given by $x_0 = (gT^2 \tan A)/4\pi^2$. Using $T = 1$ second, $x_0$ in meters is given by $x_0 = 0.248 \tan A$. You can use this equation to predict the oscillation amplitude $x_0$ for which the ball first strikes the side of the jar, if you know the angle $A$ through which the Ping-Pong ball is free to swing. To determine this angle, tilt the jar until the ball just makes contact with the side, and measure the angle of tilt.

## G.3. Rotating a ball inside an inverted plastic cup

### Demonstration

A small ball in an inverted plastic cup can be made to revolve faster and faster by shaking the cup at the right frequency, thereby illustrating both resonance and the principle of the synchrotron. In addition, the ball can be made to climb the steep walls when the cup is righted and shaken at a high enough frequency.

### Equipment

A transparent plastic cup and a small ball.

### Comment

You can place the inverted cup containing the ball on any flat surface, including an overhead projector if you want to show this demonstration to a large group. If you shake the cup back and forth at the right frequency *and in the right phase* with the rotating ball, the ball will rotate faster and faster (you will have to increase the frequency of your shaking to keep pace.) This simulates the operation of a synchrotron, a particular type of particle accelerator in which an applied voltage of increasing frequency accelerates the particles while their orbit radius is kept constant. (This is the opposite of a cyclotron, in which an applied voltage of constant frequency accelerates the particles in continually increasing orbits.)

For another interesting demonstration, place the ball in the cup right-side up, and shake the cup, causing the ball to rotate rapidly. If you achieve a sufficiently rapid rotation rate, the ball will literally climb the walls of the cup and fly off on a tangent. The ball is able to climb out of the cup because the walls make a small angle $A$ with the vertical. If the square of the ball's angular velocity exceeds the value $\omega_0{}^2 = (g/r)\tan A$, the ball will climb the walls, because the upward component of the normal force of the wall then exceeds the force of gravity. Therefore, for the ball to climb out of the cup you must shake it at a frequency at least as great as $f = \omega_0/2\pi$. The radius $r$ used in the preceding formula to predict the correct angular velocity equals the radius of the top of the cup minus that of the ball.

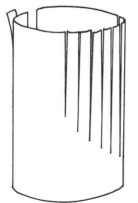

## G.4. Resonance with a slotted tin can

**Demonstration**
A "tin" can, into which a number of vertical strips of varying lengths have been cut, can be used to show resonance of one particular strip when a strip of the same length on the other side of the can is plucked.

**Equipment**
A "tin" can.

**Construction**
Using a strong pair of shears, cut about seven vertical strips in the tin can; take care not to cut yourself on the strips' sharp edges. The strips should have a width of about ½ inch and a length that varies continuously from one strip to the next. Make the longest strip about ⅔ the height of the can, and the shortest strip about ⅓ the height of the can. Be sure to cut a small gap between each pair of strips so that the strips can vibrate freely without interfering with their neigh-

bors. Also be sure to cut the strips very carefully so that they have uniform width from top to bottom. (Making them exactly equal in width is not essential.) The seven strips should take up about half the circumference of the can. On the opposite side of the can from the center of the seven strips cut another strip, whose length is identical to that of the center strip.

**Comment**

If you pluck this "exciter strip" while holding the can down, the center strip of identical length will vibrate in resonance. Other strips may also vibrate somewhat, but the ones closest in length (and natural frequency) will exhibit the largest response. If you fold the edge of the exciter strip, thereby shortening it a bit, you will raise its natural frequency slightly; this change may cause a different strip to resonate. You can also control whether the resonance is broad or narrow by making the differences in the lengths of the seven strips smaller or larger. However, for a broad resonance (smaller differences in strip lengths), cutting the strips in uniform widths is particularly critical.

The demonstration can easily be shown to a large group by putting the can on an overhead projector and focusing on the tops of the strips. (You may want to bend the strips outward a bit so they are not quite vertical and are therefore more visible when they vibrate.

## G.5. Mass on a spring

**Demonstration**

You can find the resonant frequency of a hand-held spring from which a hanging weight is suspended, by gently shaking the top of the spring at different frequencies and seeing how the amplitude varies.

**Equipment**

A fairly flexible long spring, and a weight chosen to give a suitable resonant frequency.

**Comment**

Tie the weight on the spring, or it will go flying off when you shake your hand up and down at the resonant frequency. You should try shaking your hand at a variety of frequencies greater than and less than the resonant frequency. When you are far from the resonant frequency, even large oscillations of your hand will have little effect on the oscillating weight, because the movements of your hand occur at different times during successive oscillations; sometimes these hand movements add energy to the spring-mass system and sometimes they extract energy. At or near resonance each hand movement adds energy to the spring-mass system. You can of course feel when you are close to the resonant frequency, just as you can feel at what frequency to pump a swing to get it swinging.

You may wish to repeat the demonstration using different mass-spring combinations. The easiest way to vary the spring constant $k$ is to suspend the mass $m$ from less than the full length of spring because $k$ is inversely proportional to the unstretched spring length. Given that the frequency of free oscillations is proportional to $(k/m)^{1/2}$, you should find that the frequency doubles when you either replace the mass by one that is a quarter of the original mass, or suspend the original mass using a quarter of the spring length.

## G.6. Coupled pendulums

**Demonstration**
Energy is fed back and forth between a pair of coupled pendulums at a frequency that depends on their relative lengths.

**Equipment**

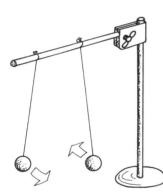

Two pendulum bobs suspended by strings, either from a horizontal rod connected to a lab stand, or to a taut horizontal string connected to fixed supports. (Note that it is important that the pendulums not be supported from a rigid structure such as the ceiling or a desk.) This demonstration can easily be done on an overhead projector if balls no larger than an inch or so in diameter are used.

**Comment**
If you use the horizontal rod suspension, be sure that the knot tying each pendulum is on the underside of the rod. If the two pendulum lengths $L_1$ and $L_2$ are identical, then when you start pendulum 1 oscillating, the small oscillations induced in the horizontal bar will begin to make the other pendulum oscillate in phase with the first.

For a more interesting demonstration, make the two lengths slightly different (say, by four percent). In this case, pendulum 2 begins to oscillate, just as before, as a result of the motion of the horizontal bar. However, since the two periods are slightly different, the oscillations of pendulum 2 gradually become out of phase with those of pendulum 1.

Pendulum 2's oscillations feed back through the horizontal bar to decrease those of pendulum 1, which eventually is brought to rest. The cycle continues, with one pendulum's oscillations alternately exciting and then diminishing the other's oscillations.

If the two lengths $L_1$ and $L_2$ differ by four percent, the two periods $T_1$ and $T_2$ (which vary as the square root of the lengths) differ by 2 percent. This means that in the time pendulum 1 takes to complete 50 oscillations, pendulum 2 completes 51. In other words, during the first 25 swings pendulum 1 is in phase with pendulum 2 (to within half a cycle) and therefore adding to pendulum 2's energy, and during the second 25 swings pendulum 1 is out of phase with pendulum 2 and therefore diminishing pendulum 2's energy and increasing its own energy in the process. Thus, the time between when pendulum 1 is most nearly at rest and when pendulum 2 is most nearly at rest is 25 times the pendulum period.

If lengths $L_1$ and $L_2$ differ by much more than four percent, you will not get good results, since the two pendulums get out of phase too quickly for the initially stationary pendulum to start swinging with a sizable amplitude. Likewise, length differences much smaller than four percent also don't give good results because you have to wait so long between reversals of the two pendulums that the oscillations damp out.

# Fluids

*Pressure and*
*Buoyancy*

### H.1. Cartesian diver

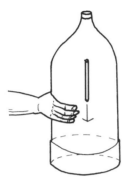

**Demonstration**
According to Archimedes' principle, a barely floating compressible object in a sealed container can be made to sink by applying a small outside pressure to the container.

**Equipment**
A plastic soda bottle, and a "diver" made from a piece of sealed soda straw that contains just enough nails to be barely buoyant. A medicine dropper partially filled with water also works well.

**Construction**
To make the diver, seal one end of a soda straw with plastic cement. Once the cement dries, drop a few small headless nails into the soda straw, and see how far the straw sinks in a glass or bottle of water. Cut the straw a short distance above the water line, and seal the top end of the straw with plastic cement, thereby creating a barely buoyant "diver." Put the diver in the plastic bottle, fill the bottle to the very top with water, and cap it.

**Comment**
When the bottle is squeezed, the pressure on the "diver" will reduce the diver's volume, causing it to displace a smaller amount of water and sink to the bottom. When you release the pressure the diver rises. The amount of pressure you must apply to make the diver sink will depend on the fraction $f$ of the straw's length that is above the water line. The diver is essentially a "threshold barometer" that sinks when the applied pressure exceeds approximately $f/(1-f)$ atmospheres. If you place two or three divers with different $f$'s in the bottle, you can see at what applied pressure each one sinks, and also measure how much pressure your hand exerts on the bottle.

If you use a plastic bottle that is only partially filled with water, you can sink a diver that is barely buoyant (small $f$), because by squeezing the bottle you increase the air pressure and decrease the volume of the diver just as before. If the demonstration is repeated using a glass bottle, however, it only works when the bottle is filled. By squeezing on a filled glass bottle, the tiny change in the bottle's volume decreases the volume of the straw by the same amount due to the incompressibility of water. On the other hand, if the glass bottle is partially filled, the diver doesn't sink, because very little change in air pressure results when you squeeze the bottle since the volume change is so small.

## H.2. Sinking a floating block

**Demonstration**
The amount of weight needed to sink a block of known weight floating in water is determined by the block's density, according to Archimedes' principle.

**Equipment**
A 2×4×4-in. wooden block; 100 pennies; and a tray in which to float the block.

**Comment**
Float the block, and add enough pennies to make its top surface just even with the water line. Probably about 50 pennies are required for the block size suggested, although the number is obviously dependent on the density of the type of wood you use. You need to distribute the pennies properly while adding them, so you don't cause the block to tip. Find the weight of the block, $W_b$, and the weight of the added pennies, $W_p$. According to Archimedes' principle, the density of the block can be expressed in terms of $W_b$, $W_p$, and the density of water $d_w$ by the equation $d = d_w W_b / (W_b + W_p)$. Calculate the density of the block and compare it to the value found by dividing the block's mass by its volume. Remember that "two-by-four" pieces of lumber are not literally 2 inches by 4 inches, so be sure to obtain the volume from a measurement of the block's dimensions.

This demonstration can also be used to illustrate the well-known puzzle of whether the water level in a pond rises or falls when you toss a heavy boulder out of a floating rowboat. When you allow all the pennies on the barely floating block to slide into the water, the water level in the container drops slightly, because the amount of water displaced by the sunken pennies is equal only to the volume of the pennies, while the amount of water displaced by the pennies on the block was equal to the weight of the pennies. The dimensions of the water tray should not be much larger than those of the block, for the drop in water level to be noticeable.

## H.3. Three holes in a water-filled bottle

**Demonstration**
The horizontal range of water spurting from a hole in a bottle is maximum when the hole is halfway between the water surface and the base, according to Bernoulli's principle.

**Equipment**
A 32-ounce plastic soda bottle; a rubber hose; and a tray with which to catch the water.

**Construction**
Put a piece of tape on the plastic bottle at the highest point at which the sides are still vertical; this height above the bottom of the bottle we will call $H$. Make three $1/16$-in.-diameter holes in the bottle, at distances $h = H/4$, $H/2$, and $3H/4$, respectively, above the base of the bottle. The holes should not be in a vertical line, but should instead be displaced by about half an inch so that the three water streams won't collide. Put the bottle on a block that is as high as the top of the tray.

An alternative—somewhat messier—way to do the demonstration, which is more suitable for a large group, is to replace the bottle by a $1/2$-in.-diameter piece of polyvinyl chloride (PVC) pipe of length $H = 6$ feet, with an end cap on the bottom, and $1/16$-in.-diameter holes drilled along its length at the $H/4$, $H/2$, and $3H/4$ points.

For either the bottle or the pipe, you need to keep adding water (using a hose or pouring by hand into a funnel) to maintain the water level at the height $H$, or else restrict your observation to the first few seconds before the water level has dropped appreciably. Also note that the demonstration will not work properly if any of the holes are clogged.

**Comment**
You should find that the water stream from the middle hole (at height $H/2$) has the greatest range of the three streams. The range is the horizontal distance traveled by the water in descending to a point even with the base of the bottle. The range depends on the product of the exit speed $v$ of the water out the hole and the time $t$ the water takes to descend. The middle hole has the maximum range because $v$ increases with water depth (and water pressure) while $t$ decreases with

water depth, so their product is highest at the halfway point, a result that can be rigorously derived using calculus. Theoretically, the range for the middle hole should be $H$, and the other two holes should have a range of $0.866H$, 13 percent less—an easily observed difference.

## H.4. Speed variation of water flowing out of a hole in a can

### Demonstration
The horizontal range of the water stream from a hole near the bottom of a can decreases linearly with time as the can empties, according to Bernoulli's principle.

### Equipment
A meterstick and two "tin" cans (12-oz or 16-oz size), one of which has an ⅛-in. hole drilled in its side very close to the bottom. The can without the hole is used to catch the water that flows out of the other can's hole. You could use a large Styrofoam cup for the "catcher," but do not use a Styrofoam cup for the container with the hole, since this container must have parallel sides, and the hole must have no ragged edges.

### Comment
When water flows out of a hole a depth $h$ below the water surface, its speed, and therefore its horizontal range, decreases at a constant rate as the can empties. The rate of decrease is constant because the speed varies as $h^{1/2}$, and it can easily be shown that $h^{1/2}$ varies linearly with time, based on the equality between the volume drop in water level and the volume of flow out the hole (the equation of continuity).

You can observe the constant rate of decrease in horizontal range by a simple experiment. First place a water-filled can at the edge of a table, with the hole (covered by your finger) slightly overhanging the edge of the table. A little water leaking from the hole will mark the point on the floor directly beneath the hole. Place a meterstick on the floor, with the 0-cm end located at this point. Orient the meterstick so that its direction parallels the direction the water stream will flow out of the can on the table. Place the "catcher" can on the floor next to the meterstick at a point where it will catch water flowing out of the first can. (You will need to uncover the hole momentarily to locate this point.) Refill the first can if the water level has dropped appreciably. Have an assistant slide the catcher can along the meterstick so that it catches the water stream as its range slowly decreases. You should record the position of the middle of the catcher can at a sequence of equally-spaced times, such as 0, 10, 20, . . . seconds. (The time $t = 0$ seconds is the instant you uncover the hole in the water-filled can.)

Make a graph of the position of the can at the recorded times, showing how the range $x$ of the water stream varies with time $t$. By eye, draw the best-fitting straight line through the data points. Read the coordinates of the points at which the straight line crosses the $x$ and $t$ axes. These axis-intercepts correspond to the initial range $x_0$ and the time required for the filled can to empty, $T$.

Compare the values read from the graph with the theoretical predictions: $x_0 = 2(Hh_0)^{1/2}$ and $T = (D/d)^2(2h_0/g)^{1/2}$. In these equations, $g$ is the acceleration due to gravity, $H$ is the height of the hole above the floor, $D$ is the diameter of the can, and $d$ is the diameter of the hole. The equations given for $x_0$ and $T$ can be derived from Bernoulli's equation and the equation of continuity. You may find that the values of $T$ obtained from the above formula and from the graph differ appreciably. If so, the most likely reason is that the hole diameter is not exactly $1/8$ inch. For example, an error of only 0.01 inch (8 percent) will result in a 16 percent error in $T$.

## H.5. Squirting a jet out of a water-filled plastic bottle

### Demonstration
Bernoulli's principle can be used to find the pressure with which you must squeeze a plastic bottle to achieve a particular maximum range.

### Equipment
A squeezable plastic bottle. Shampoo bottles are particularly suitable since they are squeezable and have a small hole in their cap. You will probably want to fill the bottle with water rather than shampoo. A water gun would be a suitable alternative to a shampoo bottle; the velocity of water exiting a water gun probably varies less from squirt to squirt than the water from a shampoo bottle.

### Comment
If the bottle is aimed directly upward the water jet will probably hit the ceiling when you squeeze the bottle (unless you

are a 97-lb weakling or the room has a very high ceiling). The easiest way to determine the jet's velocity would be to observe its range $R$ when you squirt the water horizontally from a height $H$ above the floor. The speed of the jet is then given by $v = (gR^2/2H)^{1/2}$, which follows from the equations $v = R/T$ and $T = (2H/g)^{1/2}$.

Using the computed value of $v$, you can also use Bernoulli's principle to calculate the applied pressure $P$ that your squeeze exerts on the bottle from the equation $P = \frac{1}{2}dv^2$, where $d$ is the density of water. You can also calculate the force your hand exerts on the bottle from the relation $F = pA$, which requires you to make an estimate of $A$, the surface area of that part of the bottle with which your fingers are in contact.

Another demonstration you can try is to squirt the water jet at different angles to see which angle gives you the maximum range, assuming you are able to give the bottle a squeeze with roughly the same force each time. Frequent refilling will help to keep the applied pressure constant. It would be best to squirt the bottle from floor or table height when measuring the water jet's range. You should, of course, find that the range is a maximum for an angle of 45°, and that any pair of angles above and below 45° by a given amount have the same range. Of course, the best way to produce a water stream that has a constant initial speed is to use a hose connected to a faucet.

## H.6. Pulling two plungers apart

### Demonstration
The difficulty you have in pulling two plungers apart shows how great a force atmospheric pressure can exert.

### Equipment
Two large hemispherically shaped plungers.

### Comment
Press the plungers together and get the best seal you can. If each plunger has a radius $R$, and you have a perfect seal, the force of the atmosphere on the outside of each plunger is given by the product of the projected area ($\pi R^2$) and atmospheric pressure (14.7 lb/in.$^2$). For a 6-in.-diameter plunger, this force would be 415 lb. The force actually required to pull the plungers apart is far less than 415 lb, because the interior pressure is not zero. More importantly, it is difficult to get a perfect seal, and a slight sideways force can break

the seal. Nevertheless, you may not be able to separate the plungers when holding one in each hand—at least, not until you've been pulling for a few seconds, which weakens the seal and allows some air to enter.

You may find that better results are obtained using a small suction cup from a toy arrow instead of a plunger. Moisten the suction cup, and see how much force $F$ is required to pull the cup off a smooth surface (by pulling it straight off using an attached scale). Divide $F$ by the circular area of the suction cup, and see how close the result is to atmospheric pressure.

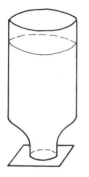

## H.7. Card under a water-filled bottle

**Demonstration**
The extent to which a card can keep water in an inverted glass depends on the height of the water and the card's flexibility, as explained by Boyle's law.

**Equipment**
A glass; a 32-oz bottle with a wide mouth; a 4×4-in. card; a 4×4-in. piece of aluminum sheet metal; and a tray in which to catch the water.

**Comment**
Most people have seen this demonstration done using a card under an inverted glass of water, but it works just as well using a 32-oz wide-mouth bottle, despite the greater weight of water resting on the card. When the card is placed underneath the inverted bottle, you will find that the water stays in the bottle whether the bottle is full, half empty, or almost empty. But if the card is replaced by the piece of aluminum sheet metal, the water stays in the bottle only when the bottle is full, even though the card and sheet have a comparable weight.

The difference in flexibility between the card and the aluminum sheet accounts for the difference, since the card can bulge more than the aluminum sheet and thus cause a greater change in the volume of the air in the bottle. According to Boyle's law any change in the air's volume must be accompanied by a reciprocal change in pressure. It may be surprising that a change in volume as small as that obtained from a bulge in the card is sufficient to cause the necessary pressure drop, but remember that to support 4 inches of water (a pressure of 0.01 atmospheres) you need a pressure reduction—

and associated volume increase—of only 1 percent, according to Boyle's law. Such a volume change is obviously easier to achieve when the bottle is nearly full than when it is half empty, since a larger increase in air volume is needed in a half-empty bottle to get the same *percentage* increase. However, the card apparently is sufficiently flexible to work even for the half-full bottle. The less flexible aluminum sheet is incapable of keeping the water from spilling out when the bottle is half full.

The primary source of the increase in air volume when the aluminum sheet is used is not the sheet's negligible bulge, but rather the slight descent of all the water in the bottle; surface tension prevents the water from pouring out the slight opening that is created. The existence of this slight gap between the aluminum sheet and the bottle can easily be confirmed by visual inspection and by the fact that the sheet can be made to slide and rattle around a bit, without spilling any water, when the bottle is inverted. Clearly, the sheet cannot descend very far before the water spills out, so it is only possible to achieve the required *percentage* change in air volume when the bottle is nearly full.

The role played by surface tension in keeping the water in the bottle can be demonstrated in an alternative amusing demonstration. Cut a circular piece of fine wire mesh to fit the mouth of the bottle, and glue it just below the opening. Despite the presence of the screen, water can easily be poured in or out of the bottle, so its presence should not be noticed by spectators. After you add some water to the bottle, put a card over the opening, and invert the bottle. Observers may not be surprised to see the card remain in place, but they certainly will be shocked to see the water remain in the bottle when you "accidentally" knock off the card. (The water does, however, pour out if you tip the bottle, so be sure to keep it vertical. Tilting the bottle causes a pressure difference across the opening, which allows air to enter one side while water pours out the other side.)

## H.8. Hanging a weight from a helium-filled balloon

**Demonstration**

By hanging a weight of the appropriate amount from a helium-filled balloon, you can achieve neutral buoyancy, thereby testing Archimedes' principle.

**Equipment**
A large balloon; some helium (available from Toys-R-Us, for example); and some weights to hang on the balloon. If the balloon is inflated to no more than 6 to 8 inches in diameter, the appropriate weight would be obtained by hanging a Styrofoam cup, to which you could add pieces of paper.

**Comment**
According to Archimedes' principle, the largest mass a helium-filled balloon can lift can be expressed in terms of the volume of the balloon $V$ and the densities of air ($d_{air} = 1.20$ kg/m$^3$) and helium ($d_{He} = 0.18$ kg/m$^3$) as $m = (d_{air} - d_{He})V$, where the mass $m$ includes the mass of the balloon itself. You can verify this relationship by adding weights to a balloon until the balloon has neutral buoyancy (goes neither up nor down).

To obtain the volume of the balloon, assume that it is a sphere, and measure its average radius. Perhaps the easiest way to find the average radius is to wrap a string around the balloon several different ways and divide the average circumference by $2\pi$. Because helium leaks out of some rubber balloons, if you let the helium-filled balloon sit overnight it will not give the same result the next day. Similarly, if you take the balloon outside, the temperature change will affect the helium density and the volume.

## H.9. Lowering a weight into a liquid

**Demonstration**
When a weight hanging from a scale is lowered into a liquid, the scale reading is reduced by the buoyant force, allowing a test of Archimedes' principle.

**Equipment**
A large demonstration scale; a weight; and a container partially filled with water. Use a weight made out of aluminum—or some other light metal—rather than steel, to get the largest possible percentage change in scale reading. You should choose a weight and scale combination that gives as full-scale a reading as possible.

**Comment**
You should observe the scale reading drop continuously as the weight enters the water. Based on Archimedes' principle, the buoyant force (the drop in scale reading as the weight is submerged) divided into the original scale reading

should equal the specific gravity of the weight. Be sure that the weight is completely submerged, but not resting on the bottom.

*Bernoulli's*

*Principle*

## H.10. Ping-Pong ball near a water stream

**Demonstration**
A Ping-Pong ball glued onto a piece of string is attracted to a stream of water from a faucet, according to Bernoulli's principle.

**Equipment**
A Ping-Pong ball glued to the end of a piece of string, and water from a faucet (or water poured out of a bottle).

**Comment**
According to Bernoulli's principle, a water stream traveling at a high velocity creates a region of low pressure. The Ping-Pong ball moves toward the water stream when placed next to it, since the other side of the ball is acted on by full atmospheric pressure. The ball does not move into the center of the stream because of the force of the descending water.

## H.11. Ping-Pong ball in an inverted funnel

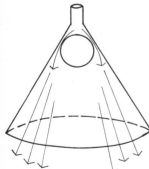

**Demonstration**
Blowing with sufficient speed through an inverted funnel that contains a Ping-Pong ball causes the ball not to fall, according to Bernoulli's principle.

**Equipment**
A funnel (preferably transparent) and a Ping-Pong ball.

**Comment**
You may first want to demonstrate that the ball can be held in place by sucking on the funnel stem, and then show that blowing also keeps the ball from falling, if the ball has been placed near the top of the funnel. The high-velocity airflow past the top right and top left sides of the ball (see illustration) causes a pressure drop at these points, while the region under the ball, where there is no airflow, is at atmospheric pressure. The resultant of these forces is upward and counteracts the weight of the ball; thus, the ball doesn't fall. For the demonstration to work, a high-speed airflow is needed to

create the necessary pressure drop. Clearly, most people couldn't blow fast enough to keep a ball much heavier than a Ping-Pong ball from falling.

For an alternative demonstration, remove the Ping-Pong ball from the funnel and place it above a vertical soda straw while you blow directly upward. If you blow hard and steady, you may be able to balance the ball in the airstream for a few seconds. In this case, the direct force of the airflow is what counteracts gravity, but the stability of the ball is explained by Bernoulli's principle. For example, if the ball begins to move out of the airstream to the left, the high-speed airflow on its right side causes a low pressure region there, so the ball is nudged back into the airstream by the higher pressure on its left side.

The demonstration is more effective if you use a hair dryer or other blower to generate the airstream. In this case, the ball—surprisingly—stays in the airstream even when the stream makes a sizable angle with the vertical. If the ball has markings, its rapid rotation will be evident as the angle with the vertical is increased. This rotation occurs because of the different airstream speeds on the two sides of the ball. For large angles the entire airstream will be on one side of the ball, which is held in place by the Bernoulli pressure drop. Fairly heavy objects such as screwdrivers can be floated in the airstream from a powerful blower.

## H.12. Blowing a quarter into a cup

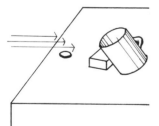

**Demonstration**
By blowing over the surface of a quarter, you can get it to jump into a cup, illustrating Bernoulli's principle and allowing you to calculate the minimum speed of your breath.

**Equipment**
A coffee mug and a quarter.

**Comment**
Set the quarter on a desk or table about an inch from the edge. Prop up the brim of the coffee mug by about an inch and place it about an inch behind the quarter (see illustration). Put your face level with the desk so that you blow horizontally over the top of the quarter. A quick hard blow should cause the quarter to jump into the cup, although repeated attempts may be necessary.

According to Bernoulli's principle, the pressure difference between the top and bottom of the coin is given by

$p_1 - p_2 = \frac{1}{2}dv^2$, where $d$ is the density of air ($1.20$ kg/m$^3$) and $v$ is the speed of your breath. If the coin barely jumps, the upward force acting on it must equal its weight, so that $F = \frac{1}{2}dv^2A = mg$. The minimum speed $v$ that will cause the coin to jump is determined from the preceding equation, which yields $v = (2mg/dA)^{\frac{1}{2}}$. For a quarter, this speed turns out to be 14 m/s. Surprisingly, the various U.S. coins require almost the same speeds, as indicated by the following table. So if you can't blow hard enough to get the quarter to jump, you probably won't have much luck with a dime either. The dime's much lighter mass is largely offset by its smaller surface area.

| | Mass (g) | Surface area (cm$^2$) | Required velocity (m/s) |
|---|---|---|---|
| penny | 2.8 | 2.69 | 13 |
| nickel | 5.1 | 3.30 | 15 |
| dime | 2.4 | 2.40 | 12 |
| quarter | 5.8 | 4.34 | 14 |

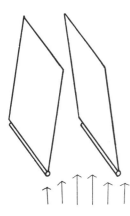

## H.13. Fan below two vertical pieces of paper

**Demonstration**
The stability of two sheets of paper, between which an upward breeze blows, is a consequence of Bernoulli's principle.

**Equipment**
A small portable fan; and two pieces of 8½×11-in. paper taped to pencils, making two paper flags. (The pencil ends of the flags are held near the fan, as shown in the illustration.)

**Comment**
The breeze from the fan blowing between the two sheets of paper creates a low-pressure region to which the papers are attracted. Hence, the space between the tops of the sheets is narrower than that between the bottoms, as shown in the illustration. (The breeze between the papers is faster than that on either side because of the narrowing of the space between the papers at their tops.) The papers are stable, because if their tops approach too closely, the direct force of the airflow will push them back.

One sheet of paper is much less stable than two, and it has a tendency to flap in the breeze like a flag. However, if you make the fan's breeze blow in the horizontal direction, and hold the piece of paper near the bottom portion of the air-flow, the sheet will assume a stable horizontal position with little flapping. The high-velocity airflow over the top side of the paper creates a lift force, just as in the case of an airplane wing. You can show that the paper's horizontal orientation is due to Bernoulli's principle, and not to the force of air hitting the underside of the sheet, by folding the end of the paper that is near the fan downward to block the airflow on the underside. This part of the demonstration can be given without a fan—simply blow hard over the top of a hand-held piece of paper and it will rise into the horizontal airstream.

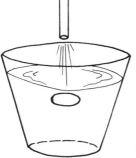

## H.14. Vortex rings

**Demonstration**
The stability of vortex rings in a fluid is a consequence of Bernoulli's principle.

**Equipment**
A clear plastic cup; a straw; and some ink, food coloring, or any colored liquid, such as soda.

**Comment**
Fluid circulations (vortices), such as cyclones, tornados, and whirlpools, tend to maintain their shape, because near the center they have higher speeds and therefore lower pressures, according to Bernoulli's principle. The vortex ring, another type of circulation pattern, has a toroidal geometry, in which the circulation around the ring produces a low pressure inside it rather than near the center, but the source of its stability is again Bernoulli's principle.

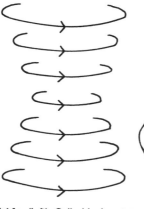
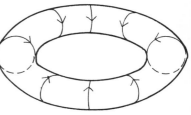

**H.14a** (left) Cylindrical vortex        **H.14b** (right) Toroidal vortex

**Fluids: Bernoulli's Principle**

To illustrate the stability of a vortex ring, fill a clear plastic cup with water and place it on an overhead projector. Place a straw in the colored liquid, and suck on the straw to fill it. Keep the liquid in the straw by putting your index finger over its top. Place the open end of the straw about an inch above the water in the cup and, keeping the straw vertical, give the middle of the straw a sharp squeeze, expelling some colored liquid into the water. A vortex should form in the water and move downward, where it will collide with the bottom of the cup and dissipate. Some practice may be necessary to get good results, as is the case with blowing smoke rings—another example of vortices.

Vortices need not be of the cylindrical or toroidal variety. For example, hurricanes combine both types of air circulation. Warm ocean water below the eye of the hurricane sup-

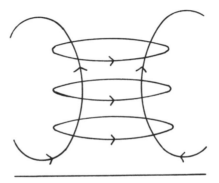

**H.14c** Combined cylindrical-toroidal vortex

plies the energy that maintains the hurricane. The warm water below the eye heats the air, giving it energy and creating an updraft at the center. Bernoulli's principle requires that the air moving upward at the center have a pressure lower than atmospheric pressure. This central pressure drop (combined with full atmospheric pressure on the outside) is the source of the centripetal force that keeps the air mass moving in a horizontal circle. When the rotating air mass loses energy because of friction, and it begins to spread out, the low pressure due to the updraft at the center pulls the rotating air mass inward and restores its rotational kinetic energy.

## H.15. Straw atomizer

**Demonstration**
An atomizer made from a plastic straw can be used to measure the speed of your breath, based on Bernoulli's principle.

**Equipment**
A plastic soda straw cut into two pieces of unequal length, and a water-filled cup.

**Comment**
Hold the longer piece of straw upright in the water with the same hand that holds the cup. Blow through the shorter piece of straw as hard as you can while holding it horizontal, a small distance above and behind the vertical piece of straw. If you are unable to produce a spray when blowing as hard as you can, try reducing the distance $H$ the vertical piece rises above the water surface. Next try varying this distance $H$ to find the largest value at which you can still produce a spray when blowing as hard as you can.

Using Bernoulli's Principle, the pressure drop at the top of the vertical straw must equal the hydrostatic pressure under a water column of height $H$. Thus, $\frac{1}{2}d_a v^2 = d_w g H$, where $d_a$ is the density of air (1.20 kg/m$^3$) and $d_w$ is the density of water (1000 kg/m$^3$). We can solve for the air velocity to obtain $v = (2d_w g H/d_a)^{1/2} = 12.8H^{1/2}$, with $H$ in centimeters. Blowing as hard as I could, I found that I was able to achieve a spray for values of $H$ up to 3.8 cm. Using the preceding equation, this yields 25 m/s for the speed of my breath.

*Surface Tension*

## H.16. Dropping pennies into a water-filled cup

**Demonstration**
The high surface tension of water can be shown by the large number of pennies that can be dropped into a "filled" cup of water before it overflows.

**Equipment**
A supply of pennies (about 100), and a cup—preferably a clear plastic cup if you want to do this demonstration on an overhead projector.

**Comment**
Because of water's high surface tension, the water surface can develop a rather large bulge before it flows over the edge of the cup. (I was able to add 66 pennies to a "filled" cup of water before it overflowed.) If you are doing this demonstration in front of a group, you might want to ask everyone to predict the number of pennies beforehand.

# H.17. Water bridge

**Demonstration**
Three liquids of different densities can be used to create an unusual configuration in which a bridge forms between the upper and lower layers—an illustration of surface tension.

**Equipment**
Rubbing alcohol, corn oil, water, and a plastic cup.

**Comment**
Pour oil into the cup, and add the alcohol to the oil. The alcohol will float on top of the oil. Now pour some water into the center of the cup. The water will descend to the bottom, usually creating a hole in the middle (oil) layer; this hole will connect the upper (alcohol) layer with the lower (water) layer through the middle (oil) layer. (It may take quite a while for the bubbles that obscure the hole to disappear.)

The "water bridge" between the top and bottom layers is stable because it is a minimum-energy configuration, which takes into account both gravitational potential energy and the energy in the stretched membrane between the liquid surfaces. For example, a slight narrowing of the hole would lead to an increase in the energy associated with the stretched surface membrane, since its surface area would increase. On the other hand, a widening of the hole would lead to an increase in the gravitational potential energy because some oil in the middle layer would have to displace some (lower-density) alcohol in the upper layer. Although the water bridge is stable for small perturbations, it can be destroyed if the liquid is stirred vigorously or heated.

Another interesting demonstration of surface tension can be made by adding a small amount of oil to an alcohol/water mixture. In this case the oil should form a large spherical globule that floats at the depth at which its density coincides with that of the water/alcohol mixture. If more oil is added, it should coalesce with the existing globule and form a larger one.

# H.18. Pepper on a water surface

**Demonstration**
When a drop of liquid soap is added to a cup of water on which pepper has been sprinkled, the resulting decrease in

surface tension causes the floating pepper to flee to the sides of the cup.

### Equipment
A plastic cup, liquid soap, and pepper.

### Comment
Place the water-filled plastic cup on an overhead projector. Sprinkle pepper on the water, and focus the projector on the pepper. When a drop of liquid soap is added to the center of the cup, the pepper suddenly flees to the edge. The pepper's motion is a result of the lower surface tension of the soap-covered water. The water surface before the soap is added can be thought of as a stretched membrane. Adding a drop of soap to the center is akin to suddenly weakening the center of a stretched membrane, which causes the formation of a hole that expands and takes the pepper with it.

---

# Heat, Thermodynamics, and Kinetic Theory

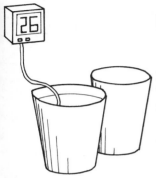

## I.1. Measuring specific heat by the method of mixtures

**Demonstration**
The specific heat of a metal block can be measured by the method of mixtures.

**Equipment**
A block made of aluminum (or some other metal); two Styrofoam cups; a thermometer with an external probe and large LCD display (obtainable for about $20 from Radio Shack, for example); an immersion heater of the kind used to boil a cup of water; and a pair of tweezers or tongs to handle a very hot piece of metal.

**Comment**
Choose the mass of the metal block so that the block, when combined with an equal mass of water, nearly fills the Styrofoam cup. Make a mark on the inside of the first cup showing the height of the water level when the cup is filled only with the amount of water whose mass equals that of the block. Before the demonstration, fill the first cup with the proper amount of water, using cold water from a drinking fountain, and fill the second cup—in which the metal block is placed—nearly to the brim with water. Heat the water containing the metal block with the immersion heater. Allow the water to boil for a minute and then quickly transfer the metal block to the first cup, which contains an equal mass of cold water at a temperature $T_c$. Continuously record the temperature of the water and note when it reaches a maximum value. This is the equilibrium temperature of the mixture, $T$. The heat gained by the water, $m(T - T_c)$, equals the heat lost by the metal, $mc(100 - T)$. Therefore, the specific heat of the metal can be found using the equation $c = (T - T_c)/(100 - T)$.

For best results, it is important that $T_c$ be below room temperature by roughly one tenth the number of degrees that $100°C$ is above room temperature. If this is the case, as long as the specific heat $c$ is around 0.1 cal/(g·°C), the final equilibrium temperature $T$ will be close to room temperature. This is important, since otherwise the mixture can lose a significant amount of heat during its approach to equilibrium. The best way to ensure that the water is at the proper initial temperature is to get cold water from a drinking fountain, and use your immersion heater to heat it to the optimum temperature before you begin the demonstration. Remember that immersion heaters should not be plugged in unless they are completely submerged.

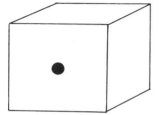

## I.2. Hole in a box

**Demonstration**
The blackness of a hole in a box that has a white interior serves as a good example of a "blackbody," or perfect absorber.

**Equipment**
An enclosed cardboard box with a hole cut in the front. The diameter of the hole should be about one tenth the shortest dimension of the box. You may wish to prepare two boxes, one with a white and one with a black interior, that can easily be opened to show that the hole's blackness is unaffected by the color of the interior.

**Comment**
The blackness of the hole is due to the fact that only a tiny fraction of the light that enters the hole comes back out, since most of it undergoes multiple diffuse reflections and is absorbed by the inside walls before getting back to the hole. It is surprising how large the hole can be and still appear reasonably black. You may wish to experiment using holes of different sizes in boxes with white and with black interiors. One way to use just a single pair of boxes would be to make a large hole in each box, and cover up the hole with cards that have holes of different sizes.

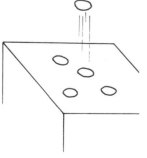

## I.3. Bimetallic jumping disks

**Demonstration**
Differential expansion due to heating, and the transformation of elastic potential energy into kinetic energy, are illustrated using "jumping disks."

**Heat and
Thermodynamics**

**Equipment**
"Jumping disks," sold by the Edmund Scientific Company.

**Comment**
The disks are made by bonding together flat sheets of two metals with different thermal expansion coefficients. Due to differential thermal expansion, the equilibrium curvature of the disk, as well as the forces present at a given curvature, depend on temperature. At some temperatures, curvature of one sign (concave-up or concave-down) is the stable configuration, while at other temperatures curvature of the opposite sign is stable. Suppose the top surface of the disk has the higher thermal expansion coefficient. Clearly, concave-down would then be the stable curvature for high temperatures, and concave-up for low temperatures—"high" and "low" temperatures, that is, relative to the temperature at which the two metals were bonded into a flat sheet (75 °F). Obviously, if you start with a hot disk concave-down and let it cool, at some temperature it will pop into the concave-up position. The forces generated when it pops can cause it to jump quite high. Depending on how warm your hands are, hand-warming may not be enough to make the disks stay "clicked in" the concave-down state. Putting the disks on an overhead projector is an effective way to warm them up (as well as show when they jump).

The disks work best at temperatures between 72 °F and 78 °F (a symmetrical range above and below the bonding temperature of 75°). Below 70° it may be difficult to put a disk into the "clicked-in" position. At 80° and above, there may be a long delay before the disk jumps, or the disk may not jump at all. A bunch of disks that all clicked-in at the same time will jump at random times.

Owing to the random jump times, a collection of $N$ disks on an overhead projector roughly simulates the radioactive decay of $N$ nuclei. The decay "half-life" can be found by measuring the time it takes for half the disks to jump after the projector is turned off and the disks are allowed to cool. Unlike the decay of atomic nuclei, the disks' half-life is obviously affected by the temperature of the environment. Another fundamental difference is that the disks do not satisfy the exponential-decay law, because although their "decays" are random in time, each disk does have a "memory" of how long it has been warming up—unlike nuclei, which have no such "memory."

We might expect the "decay times" for a large number of disks to follow a Poisson distribution, which in principle could be tested by direct observation. However, it would be far easier to determine the mean decay time and the standard deviation about the mean, because the number of decays needed to find these two parameters reliably is far fewer than the number needed to find the shape of the distribution.

## I.4. Molecular motion simulation

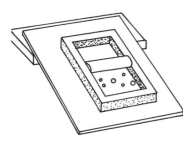

**Demonstration**
The dependence of a gas' volume on its pressure and temperature can be modeled using BB's bouncing in a hand-agitated frame with a movable piston.

**Equipment**
A 4×6-in. frame made from ½×½-in. pieces of wood (obtainable at a crafts store); a "piston" made from a ¾-in.-diameter dowel; some BB's (both ⅛-in. and ¼-in. sizes); and a transparent 8×10-in. surface (a sheet of either acrylic or glass). The BB's can be obtained at a gun or sporting-goods store.

**Construction**
Purchase a small wooden picture frame, or make the 4×6-in. wooden frame by gluing the ½×½-in. pieces of wood together. Be sure that the sides of the frame are as parallel as you can make them. Cut a segment of the ¾-in.-diameter dowel (the "piston") that just fits within the frame. When the frame is placed on a flat horizontal surface, the dowel should be able to roll freely without snagging anywhere along the inside of the frame, but with as little clearance as possible between itself and the frame. Now put the frame and dowel on the transparent acrylic sheet, and place the entire apparatus on an overhead projector. The top edge of the acrylic sheet should be propped up just enough that the dowel always rolls down when it is placed anywhere within the frame. (If you left too much clearance between the dowel and the sides of the frame, the dowel may tilt a bit and "hang up.")

**Comment**
Place some large (¼-in.) BB's inside the frame below the "piston" and start shaking the frame by hand. If you shake hard enough the impact of the BB's will drive the piston upward. This simulation illustrates the statistical nature of

pressure. It also illustrates the fact that—for a constant pressure—the higher the temperature (the more vigorously you shake the frame), the more the "gas" expands.

Next, you can vary the "piston" pressure by tilting the acrylic surface at a larger angle. You will find that you need to shake more vigorously (produce a higher temperature) to keep the piston at the same height it was before. Finally, remove the piston and add both large and small BB's (about 10 of each). When you shake the frame, the heavy BB's don't move as fast as the light BB's, thereby illustrating the fact that for a mixture of two gases at a common temperature, the heavier molecules must have lower velocities. (You need enough BB's so that enough collisions between BB's occur to give the two sizes a common "temperature.")

For still another demonstration, replace the frame with one that has a wall dividing the frame into two equal compartments. The wall should have a half-inch opening that allows BB's to pass from one side to the other when the frame is shaken. A demonstration of the entropy law starts with some BB's on one side of the wall and none on the other. As the frame is shaken, BB's passing through the opening at random tend to equalize the number on each side as time goes on. However, owing to the random nature of the process, the side with more BB's will occasionally gain rather than lose a BB. On a percentage basis such fluctuations are more probable for small numbers of BB's, which models the statistical nature of the approach to equilibrium, and its occasional fluctuations, for isolated systems.

An interesting demonstration would be to time how long it takes for random shaking of the frame to result in all the BB's being in the right compartment, if all $N$ start out in the left compartment. On the average, the time for this to occur should be $2^{N-1}$ times the average time a single BB takes to move from one compartment to the other.

## I.5. Radiometer

**Demonstration**
The turning vanes of a radiometer placed in a bright light illustrate the pressure due to molecular recoil.

**Equipment**
A radiometer (also called a pyrometer) and a light source. (If no light source is handy you can use the light from an overhead projector turned on its side.)

**Comment**

The radiometer, which has vanes that are mirrored on one side and black on the other, has been evacuated to a pressure of approximately $10^{-2}$ mm of mercury. The mean free path of gas molecules at this pressure is about the size of the radiometer bulb, which allows for a particularly efficient transfer of momentum between the molecules and the vanes. When the device is exposed to a bright light, molecules incident on the hot black side gain more momentum than those incident on the cooler mirrored side, causing the black side to recoil.

An incorrect explanation often given for the rotation of the vanes is based on the different pressure of light photons on the black and mirrored sides. But this would require the vanes to rotate in the direction opposite to that actually observed, since the momentum imparted by rebounding light photons striking the mirrored side is twice that imparted by absorbed photons striking the blackened side. In fact, this effect is negligible, because although the photons carry a lot of energy, they carry little momentum compared to the gas molecules. Photons carry little momentum for their energy compared to molecules because the momentum-to-energy ratio of photons is $1/c$ (where $c$ is the speed of light), while for molecules of velocity $v$ much less than $c$, the ratio is $2/v$.

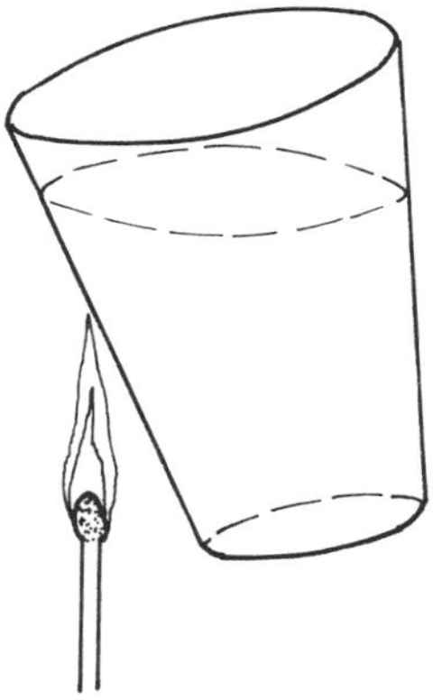

## I.6. Burning paper and Styrofoam cups

**Demonstration**

Whether or not a paper cup will burn when a lit match is applied to it depends on the thermal properties of the cup and its contents.

**Equipment**

A wax-covered paper cup; a Styrofoam cup; sand or dirt; copper shot; and a solid metal cylinder 1 inch or more in diameter.

**Comment**

Whether or not a cup burns depends on the thermal conductivity of the cup and its contents, as the following observations confirm:

**1.** If you apply a lit match to the bottom of a water-filled paper cup, the bottom ledge and some of the wax on the sides burns, but the cup itself does not, even if the water is brought to a boil. The high thermal conductivity of the thin paper allows the heat from the match to flow through readily, so the outside of the cup cannot get much hotter than the inside, which in the case of boiling water is at a temperature of 100 °C—not hot enough to ignite the paper.

**2.** If you apply a lit match to the bottom of a water-filled Styrofoam cup, you can burn a hole in it, causing the water to spill out. The Styrofoam cup is a better insulator (poorer conductor) than the paper cup, so a larger temperature difference can exist between the outside and the inside. Applying the lit match causes the cup's exterior to reach the kindling point, even though the inside of the cup cannot get hotter than 100 °C.

**3.** If you apply a lit match to the bottom of a sand-filled paper cup, the cup burns, although it may remain intact. (You can show that the cup burns through by dumping out the sand to see the blackened interior.) Unlike the water-filled paper cup, for which the outside could not get much hotter than the 100 °C temperature of the inside, the temperature of the sand-filled cup has no such limit. In addition, since sand is a poor conductor, heat that builds up in the paper cannot readily dissipate through the sand.

**4.** If you apply a lit match to the bottom of a paper cup filled with copper shot, it burns (and will create a mess unless the shot is caught in a tray). Despite the good thermal conductivity of copper, the granular nature of the shot results in poor thermal contact with the wall of the paper cup. Heat, therefore, can build up in the paper without dissipating into the copper, and the paper burns.

**5.** If you apply a lit match to a piece of paper taped around a metal cylinder, it does not burn. Metals are excellent thermal conductors, and thus heat flowing through the sheet of paper is quickly dispersed throughout the metal. No heat buildup occurs next to the paper's interior surface, as is the case with the cup filled with copper shot, because of the good thermal contact between metal and paper.

## I.7. Heat conductivity demonstration using thermostrips

### Demonstration

The thermal conductivity of a plate can be gauged from its effect on the temperatures recorded by rows of thermal strips in contact with the plate.

**Equipment**

Ten thermostrips; a transparency blank; two 8×10-in. plates, one made of a poor thermal conductor and the other of a good thermal conductor (for example, wood and aluminum); and a tray in which to put some hot water.

**Comment**

Tape the ten thermostrips onto the transparency blank, and tape the blank onto one of the 8×10-in. plates. Stick the lower end of the plate into some hot water in the tray. For the poorly conducting plate, the temperature rise registered on the thermostrips should be confined to the lower one or two strips, while for the plate made of the good conductor, higher rows of thermostrips should register increased temperature.

## I.8. Measuring thermal conductivity from rate of heat loss

**Demonstration**

The thermal conductivity of Styrofoam can be determined by measuring the rate at which the temperature of water in a Styrofoam cup decreases.

**Equipment**

A Styrofoam cup; a thermometer, preferably a digital thermometer with external probe and large LCD display (obtainable for about $20.00 from Radio Shack, for example); and an immersion heater.

**Comment**

Using the immersion heater, heat a cup of water to the highest temperature the thermometer can read. (Some digital thermometers go up to only 50 °C.) Make a small hole in the cup's lid and place the lid on the cup, with the thermometer probe inserted in the hole. The measurement accuracy can be improved if a snugly fitting piece of Styrofoam is used instead of a plastic lid. Measure the time that it takes for the water temperature to drop one degree. Be sure that the probe is deep enough in the water that it doesn't serve as a conduit for significant heat loss. Also, be sure to allow enough time for the thermometer to "settle down," if it is digital.

The thermal conductivity of Styrofoam, $k$, is given in terms of measurable quantities by the equation $k =$

mcd/$At(T - T_0)$, where $m$ is the mass of the water in grams, $c$ is the specific heat in cal/(g·°C), $d$ is the cup's wall thickness in centimeters, $A$ is the cup's total surface area (including top and bottom), $t$ is the time in seconds for the water to cool 1 °C, and $T - T_0$ is the difference between the water's temperature and room temperature. (Since the water temperature drops by only one degree, it doesn't matter much whether you use the original temperature or the final temperature in finding $T - T_0$).

The preceding equation for $k$ assumes that the heat flow through the cup's walls is a one-dimensional problem analogous to heat flow through a flat slab; this is a good approximation as long as the cup is sufficiently well insulated to make the water temperature uniform throughout. In practice, however, finding $k$ by applying this equation to water in a cup is not terribly accurate; the main sources of error are (1) inaccuracy in determining the cup thickness $d$, (2) heat loss through the thermometer probe, (3) different heat conductivity of the plastic lid and of the Styrofoam cup, and (4) different rates of heat loss through different parts of the cup, due to temperature differentials. The expected value for the thermal conductivity of Styrofoam, $6 \times 10^{-5}$ cal/(sec·cm·°C), is essentially the same as that of air because Styrofoam consists of many separated air pockets.

## I.9. Rate of cooling

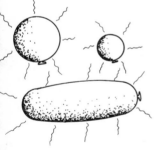

### Demonstration
The rate of temperature drop of a hot uninsulated object depends on its mass, surface area, temperature, specific heat, and surrounding medium.

### Equipment
An assortment of balloons; a large tray; and a thermometer, preferably the digital type, with an external probe and large LCD display (obtainable from Radio Shack for about $20.00). The balloon assortment should include one long cylindrical balloon and three round ones.

### Comment
Hot objects lose heat to their surroundings by the mechanisms of conduction, convection, and radiation. Radiation is important when an object is much hotter than its surroundings, or when the other two mechanisms are suppressed, as in the case of an object surrounded by vacuum. Convection

is important when a fluid circulation pattern exists or is induced by the hot object.

The rate of temperature drop of a large hot object that loses heat primarily by natural convection and radiation can be expressed as $R = kA(T - T_0)/mc$, where $k$ is a constant, $A$ is the object's surface area, $m$ is its mass, $c$ is the specific heat, and $T - T_0$ is the temperature difference between the object and its surroundings. For spherical objects of density $d$ and radius $r$, we may substitute $3/rd$ for the ratio $A/m$, and the rate of temperature drop can be expressed as $R = K(T - T_0)/r$, where $K = 3k/cd$ is a constant that depends on the material.

According to this equation (which holds only for objects moderately warmer than their surroundings), small spheres cool more rapidly than large ones. In more general terms, large objects tend to cool more slowly than small ones, assuming no significant differences in their density, shape, or specific heat. This accounts for the higher metabolisms of small mammals, who have to scurry about to get the food they need to maintain their body temperature. If we consider two objects of the same mass, density, and specific heat, the one that more closely resembles a sphere (the shape with the lowest area-to-volume ratio $A/V$) loses heat more slowly. This accounts for the tendency of small mammals to roll up into a ball to conserve their heat in cold weather. The heat conductivity of the surrounding medium is also an important factor in determining the rate of heat loss. For example, mammals lose body heat much faster in water than in air.

Using water-filled balloons, you can make some simple measurements to verify the dependence of the cooling rate $R$ on an object's size, shape, temperature, and surrounding medium:

**1.** Size and temperature. Insert the external probe of the digital thermometer in the stem of a spherical balloon, and fill the balloon with hot tap water to some specific radius. Measure the temperature at the beginning and end of a five-minute interval. Repeat the procedure using a number of spherical balloons filled with water at various initial temperatures. Use balloons of various sizes, with radius ranging from 1 to 4 in. You can determine the average radius of a balloon by measuring its average circumference with string and dividing by $2\pi$. To test the equation $R = K(T - T_0)/r$, compute the quantity $Rr/(T - T_0)$ for each balloon, and see if it is a constant, $K$, for all balloons. ($R$ is the measured temperature drop in five minutes, $r$ is the average balloon radius, and $T - T_0$ is the average amount the balloon's temperature

exceeds room temperature during the five-minute interval.)
You will probably find that the equation is valid over a range
of temperatures and balloon radii, except for very small balloons, which lose heat much faster than the equation pre-
dicts—probably because the size of the internal convection
cells changes abruptly at a certain balloon radius.

**2.** Shape. Fill a spherical balloon and a long cylindrical
balloon with hot tap water, so that their water temperature
and masses are the same. The long balloon has a larger area-
to-volume ratio than the spherical one, and hence you should
find its five-minute temperature drop is also greater.

**3.** Surrounding medium. Fill a spherical balloon with hot
tap water to some specific radius $r$, and measure its temperature at the beginning and end of a five-minute interval during which it is in a water bath at room temperature. You
should find a much larger five-minute temperature drop than
for balloons not immersed in water, because water is such
a good conductor of heat. In other words, if you compute
$Rr/(T - T_0)$ for several balloons immersed in water, the values should be roughly constant, and much greater than those
for balloons surrounded by air.

The fact that the rate of temperature drop of a hot object
is proportional to the difference between its temperature and
that of its surroundings requires that the object's temperature exponentially approach that of its surroundings—Newton's law of cooling. This law can be checked by measuring
the object's temperature at a sequence of times.

## I.10. Mechanical equivalent of heat

**Demonstration**
The number of joules equivalent to one calorie (the mechanical equivalent of heat) can be determined from the rise in
the temperature of a bag of lead shot that is repeatedly
dropped.

**Equipment**
A bag partly filled with lead shot; a thermometer.

**Comment**
The fact that heat is often produced when work is done
should be familiar to anyone who has ever used sandpaper,
extracted a nail from a board, or smelled burning rubber
when car tires skid. The relationship between mechanical
energy in joules and heat energy in calories can be found by

repeatedly dropping a bag of lead shot a known distance. If a bag containing a mass of lead shot $m$ is dropped 50 times from a two-meter height, the thermal energy increase of the shot equals the potential energy loss: $E = 50mgh = 980m$ joules. The thermal energy increase can be expressed in terms of the observed temperature rise, $\Delta T$, as $Q = 31m\ \Delta T$, where we have used the value 31 cal/(kg·°C) for the specific heat of lead. If $\Delta T$ is measured, the mechanical equivalent of heat can be found from the ratio of the mechanical energy $E$ to the heat $Q$: $E/Q = 980/(31\ \Delta T)$. For this ratio to be close to the accepted value of 4.186 joules/cal, the temperature rise of the shot must be about 7.6 °C.

Lead is a particularly good material to use in this demonstration because of its low specific heat, which leads to a large temperature rise. An equivalent experiment done with a substance that has a higher specific heat would give a smaller temperature rise—for example, 2.5 °C for copper and only 0.25 °C for water. Although the measured temperature rise is in principle independent of the mass, you don't want to use too small a mass. Small masses lose their heat too quickly, owing to their large surface-area-to-volume ratio. If you vary the mass of the shot as well as the number of times it is dropped, you will see how much measurement error can be ascribed to heat loss, and how much to other sources such as variations in temperature in different parts of the lead shot.

## I.11. Apparent violation of the entropy law

### Demonstration
A dye mixed into a glycerine solution can surprisingly be unmixed—an apparent violation of the entropy law (the second law of thermodynamics).

### Equipment
Food coloring; glycerine (obtainable at a drug store); and a small cylindrical clear plastic box (about 2 inches in diameter), with a snugly fitting cover. Plastic boxes can be obtained from plastics companies and some surplus outfits, such as Jerryco, Inc.

### Comment
The standard version of this famous demonstration involves two concentric transparent plastic cylinders, one of which can be rotated inside the other. The version described here uses much simpler apparatus to make the same point, and it

works well on an overhead projector. Place some glycerine in the plastic box, filling it to the very top. Put a large drop of food coloring at an off-center point, on or beneath the glycerine surface. Put the cover on snugly, and turn the box upside down. No glycerine should leak out if the cover fits snugly, and no bubbles should appear if the box was filled to the top. Wait a little while until parts of the food-coloring drop rise through the glycerine, forming a straight vertical line, and then place the box on an overhead projector. If you rotate the box relative to the lid, different parts of the dye will rotate by different amounts, transforming what was originally seen as a small blob on the screen into a circle as the box is rotated—apparently showing the mixing of the dye into the glycerine. If you now rotate the box in the opposite direction the dye surprisingly returns to form the original blob—an apparent example of "unmixing." Unmixing occurs, even if you rotated the box through many turns, provided not too much time has elapsed. The reverse rotation returns the dye to its original blob form because the flow is lamilar, and therefore reversible, at all times.

No violation of the entropy law is actually at work here because the spreading out of the dye from the blob does not represent any increase in entropy or disorder. In fact, a side view of the box reveals that the rotation does not cause a point-like blob to transform into a circle, but rather the rotation transforms a vertical line into a helix; no mixing actually occurs.

In the classic variant of this demonstration, a vertical line of dye is deposited in the glycerine-filled space between two concentric transparent cylinders. If the inner cylinder is rotated, the vertical line of dye is smeared into a cylindrical shell, but here again no mixing actually occurs. The original vertical line of dye has a finite thickness; molecules on the inner edge of the line rotate through angles different from those in the middle or outer edge, causing the apparent smearing. The original vertical line is restored when the rotation is reversed.

# Waves
*Traveling*

*Waves*

## J.1. Waves along a spring or rubber hose

**Demonstration**

The nature of traveling waves and the relationship between frequency, wavelength, and velocity can be shown by shaking one end of a long spring or very heavy rubber hose (for transverse waves) and a slinky (for longitudinal waves). Actually, you could use the slinky to demonstrate both longitudinal and transverse waves, depending on whether you shake one end back and forth or side to side. If one slinky is not long enough you could connect several together.

**Equipment**

A spring (at least six feet long) or a heavy rubber hose (twenty feet long) for transverse waves, and a slinky for longitudinal waves. The very heavy long rubber hose with 1-in. outer diameter and ⅜-in. inner diameter is better than the spring for transverse waves because it can be obtained in arbitrarily long lengths and shows less attenuation along its length.

**Comment**

Lay the long piece of hose in a straight line on the floor, and create a pulse in the end you are holding. Next, create a sine wave by shaking your hand vigorously in a horizontal direction. (This may not work too well if the rubber hose is on a carpet with appreciable friction, in which case you should use the spring instead.) Notice that because of friction the wave amplitude is reduced as it travels along the length of the hose. One beneficial effect of the attenuation is that you don't have to worry about waves reflected from the end of the hose, since the waves never reach the end if the hose is long. By shaking the end of the hose at different frequencies you can show the inverse relationship between frequency and wavelength for waves of fixed velocity. You can then use the slinky to demonstrate traveling longitudinal waves. Reflections from the end of the slinky can be reduced if it is shaken on a desk—this causes waves to be attenuated as they travel down the slinky.

Traveling Waves **J.2.** Rolled-up transparency for transverse waves

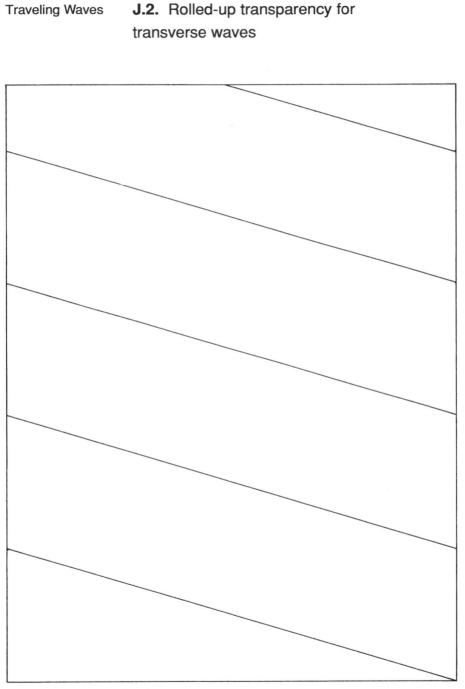

**J.2a** Transparency template for transverse waves

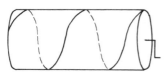

**Demonstration**

A rolled-up cylinder, made from a transparency of Plate J.2a, gives an image of a traveling sine wave on an overhead projector when the cylinder is rotated.

**Comment**

Match the left and right vertical lines when rolling up the transparency, and keep it cylindrical with transparent tape. You can make other transparencies corresponding to waves of different wavelength and amplitude by varying the number and slope of the lines. The faster you rotate the rolled-up transparency, the higher the frequency and the velocity. You may want to make a little handle out of a paper clip stuck on a cardboard end-cap to make rotation easy.

## J.3. One-dimensional water wave in a narrow trough

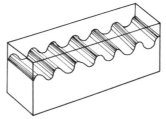

**Demonstration**

One-dimensional water waves can be observed in a long, narrow transparent trough.

**Equipment**

You can make a long, narrow trough from a sheet of ¼-in.-thick acrylic (obtainable at a hardware store or from a plastics company). Suggested dimensions are $6 \times 2 \times 40$ inches. Considerable patience and care is required to make a leak-proof trough. It may be advisable to add food coloring to the water to make the waves visible to a large group.

**Comment**

You can observe a variety of phenomena, including propagation of both wave pulses and sinusoidal waves. Wave pulses can easily be created by quickly raising and lowering one end of the trough, and sinusoidal waves can be created by allowing a mass on a spring to oscillate in and out of the water near one end. (A vibrating plastic ruler with a suspended clay ball also makes an excellent source of sinusoidal waves, as described in demonstration Q.3.)

**Standing Waves**

You should observe how the wave speed $v$ depends on both the wavelength $\lambda$ and the water depth $h$. You can estimate the speed by timing the wave as it moves the length of the trough. For water of a depth much greater than the wavelength, the wave speed depends on the wavelength, and can be shown to have a minimum value of 23.5 cm/s for a wavelength of 1.73 cm. For "shallow" water—defined as having a depth much less than the wavelength—the water speed can be shown to be $v = (gh)^{\frac{1}{2}}$, independent of wavelength.

## *Standing Waves*

## J.4. Rolled-up transparency for longitudinal waves

**Demonstration**
A standing longitudinal wave can be simulated by two cylinders made from transparencies of Plate J.4a, when one is slid inside the other.

**Comment**
Plate J.4a represents a snapshot in time showing the positions of particles in a medium through which a longitudinal wave is passing—for example, the air molecules disturbed by a sound wave. Regions of the pattern that have a high density of dots represent compressions, and regions of low density represent rarefactions. In a real sound wave, the fluctuations in density are typically far less than those depicted in Plate J.4a. If you make two transparencies from the figure, you can roll them up to form cylinders. When one cylinder is slid back and forth inside the other, you obtain the appearance of a standing wave. (At one instant compressions and rarefactions overlap, producing a uniform density of dots everywhere, and a quarter cycle later compressions overlap compressions, producing distinct maxima and minima.) The simulation is not completely accurate, because in a real standing wave, two traveling waves propagate in opposite directions, and do not oscillate back and forth. This inaccuracy cannot easily be removed. For example, if we tried to simulate two traveling waves by sliding one cylinder at a constant speed relative to the other, that would create still another misinterpretation, making it seem as if the individual molecules were moving at the same speed as the wave.

Section J

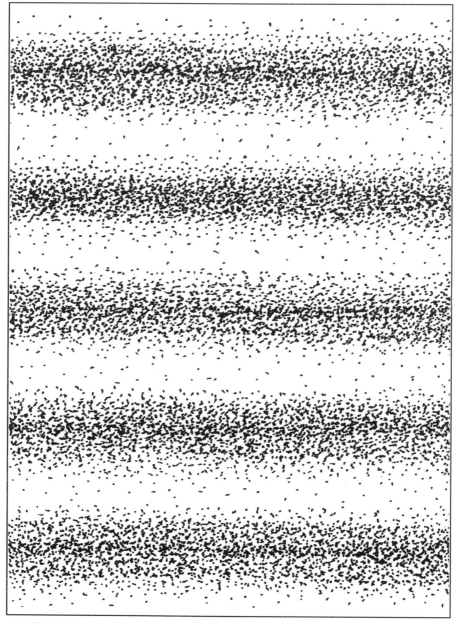

**J.4a** Transparency template for longitudinal waves

## J.5. Sound waves in a variable-length tube

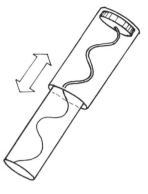

**Demonstration**

Standing sound waves can be produced using a beeper at the end of a variable-length tube, when the length is an integral number of half-wavelengths, the condition for resonance.

**Equipment**

A beeper with a sound wavelength of about four inches. A good choice is Radio Shack model 273-060A, which operates at 4 to 28 V DC and has a resonant frequency of 3600 Hz ± 500 Hz. You will also need two thin-walled hollow tubes about two inches in diameter and a foot long, one of which can slide within the other. Transparent tubes are desirable if you want to show the demonstration on an overhead projector.

If you have only one transparent plastic tube two feet long, you could cut it in half and then slit one of the halves along its length, allowing you to fit the slit half inside the other. Acrylic tubes can be ordered from plastics companies in various sizes. Another possibility is to make two tubes from rolled-up transparency blanks.

**Construction**

Tape the beeper to a 9 V battery with duct tape. Connect the beeper's leads to the battery and to a line cord, so that it can be activated by connecting the two wires at the other end of the line cord. Place the beeper and battery at the end of the inner tube, which should fit snugly inside the other tube. Make a drawing of standing waves of 4-in. wavelength and make two transparencies from the drawing. Tape one transparency onto each tube. The transparency taped onto the tube holding the beeper should have a node at the location of the beeper. The one taped onto the other tube should have an antinode at its open end.

**Comment**

Activate the beeper and vary the overall tube length by sliding one tube inside the other. Resonant standing sound waves will occur when the length of the tube equals an integral multiple of a half-wavelength. Thus for a 4-in. wavelength, the sound level should show a series of sharp maxima for every two inches the tube length is extended. If the drawing used to make the transparencies is accurate, and the wavelength is indeed four inches, you should find that the standing wave drawings on each tube smoothly join at the same time you hear the maxima. Bear in mind, however, that the beeper frequency has an uncertainty of about 14 percent, so the actual wavelength may show a similar departure from its nominal value.

## J.6. Snipping a straw as you toot

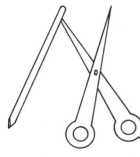

### Demonstration

Cutting pieces off the end of a straw, while you blow through it, creates a sound that becomes increasingly high pitched, as required by the resonance condition for standing waves in an open-air column. A more dramatic version of the variation of sound frequencies can be demonstrated using an oscilloscope and a microphone.

### Equipment

A soda straw and a pair of scissors. Flatten the end of the straw you will blow through by chewing on it, and cut it to resemble a bird's beak, as shown in the illustration.

### Comment

Try blowing with various amounts of force until you produce a loud toot, and then snip off the end of the straw piece by piece. The explanation for the increase in frequency heard as the straw length is decreased follows from the standing-wave condition for a tube open at one end. The fundamental wavelength is twice the tube length, so the fundamental (and higher harmonic) wavelengths decrease, and frequencies increase, as the tube is shortened.

The inverse relationship between frequency and wavelength for sound can be demonstrated in a much more dramatic fashion if you have an oscilloscope and a small microphone. Connect the microphone leads to the oscilloscope and whistle into the microphone. If you whistle at a continuously rising pitch, the trace on the screen should be approximately sinusoidal, and the wavelength should continuously decrease as the pitch rises.

## J.7. Harmonic frequencies in a corrugated tube

### Demonstration

Three or four different resonant frequencies can be excited by blowing through a corrugated tube at different speeds, illustrating a surprising interaction between a random phenomenon and an orderly one.

## Equipment
A toy known variously as a "whirl-a-tube," "bloogle," or "hummer," available from Toys-R-Us for under $2.00. The longer, flexible variety is meant to be twirled around in a circle. The shorter, narrower type, which costs about 20 cents, is blown like a whistle.

## Comment
Both types of hummers produce a sound because of airflow through the corrugated tube. The one that is whirled in a circle uses the low pressure at the whirled end—which arises from Bernoulli's principle—to create the airflow from the other end. The sound produced by the hummer has several puzzling properties, in particular the creation of certain frequencies only at certain airflow speeds, and the abrupt transitions in frequency with flow speed.

In order to understand this behavior, we must explain how sound is created in a corrugated tube. When air flows past the tube's internal corrugations, they are attracted inward by Bernoulli's principle. This inward motion of a corrugation slightly restricts the airflow, producing a pressure fluctuation that alters the way the airflow interacts with subsequent corrugations. The random pressure fluctuations constitute a sound wave that is analogous to the sound of a flag flapping in the breeze. Both the flapping flag and the vibrating corrugations vibrate at a higher frequency with increasing air speed, but they also show a randomness—making it more correct to refer to a range of frequencies than to a single frequency, which pertains only to a regular periodic motion. Thus, the range of frequencies present in the sound (the "frequency spectrum") shifts toward higher frequencies as the air speed through the tube increases.

Plate J.7a shows three possible frequency spectra corresponding to the airflow velocities $v_1$, $v_2$, and $v_3$. The sharp spikes labeled $f_1, f_2, f_3,$ and $f_4$ show the frequencies of four harmonics for standing waves in the tube. Clearly, only $f_1$ is excited at air speed $v_1$, only $f_2$ at air speed $v_2$, and only $f_3$ at air speed $v_3$. Harmonic $f_4$ is not excited until an air speed higher than $v_3$ is reached.

If the corrugations have the right size and spacing, the

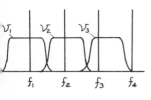

**J.7a** Frequency spectra for three airflow velocities in a corrugated tube

width of the frequency spectrum is comparable to the distance between the harmonics for standing waves in the tube. In that case, one—and only one—harmonic is excited, as an increase in the air speed causes the frequency spectrum to overlap progressively higher harmonic frequencies.

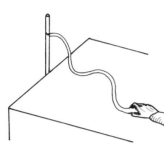

## J.8. Standing waves in a coiled spring

**Demonstration**
When you excite standing waves by shaking one end of a long spring, the frequencies of the harmonics are found to be multiples of the fundamental frequency.

**Equipment**
A 6-ft-long coiled spring; a slinky; and a metronome, if available.

**Comment**
Be sure the spring is very firmly attached to a fixed object at one end. Otherwise, it will fly off because of the large forces generated when a wave reaches the end. You might have an assistant hold one end fixed, or clamp it firmly to a vertical rod that is itself firmly clamped to a desk or table. Lift the spring off the table, and shake its free end at a frequency that will excite the fundamental (one half-wavelength along the length). Try to excite the second and third harmonics by shaking your hand up and down at two and three times the observed fundamental frequency, respectively. It is difficult to excite harmonics beyond the first few, and even the first few will require some hand-eye coordination. Essentially, you need to vary the frequency slowly, until the proper pattern begins to emerge. Perhaps the best way to excite a given harmonic would be to shake the spring in time with a metronome set at the predicted $n$th harmonic frequency, once you have determined the fundamental frequency.

Longitudinal standing waves in a slinky can be excited in a similar manner. A method that works well on an overhead projector is to use a smaller version of the slinky—the "slinky junior." First tape the two ends of the stretched slinky onto the overhead projector, and brush the tip of a pen or pencil against the slinky coils to create a wave pulse that travels down the length of the slinky and back. To create the fundamental standing-wave mode, brush a pen or pencil against the coils with a period equal to the time the wave pulse required for a round-trip. Higher harmonics can be produced by using integral multiples of this frequency.

## J.9. Longitudinal waves around a circle

**Demonstration**
Longitudinal standing waves can be created around a circle by using a small slinky wrapped around a cylinder and connected end to end.

**Equipment**
A "slinky junior," available from Toys-R-Us for about $1.00; and a 2-in.-diameter cylinder.

**Comment**
Connect the slinky end to end, using some string or wire, and place it around the cylinder on an overhead projector. Produce a wave pulse by brushing the tip of a pen or pencil against the slinky coils, and observe the time the wave requires to travel around the slinky. Now brush the pen or pencil tip back and forth against the coils with a period that approximates that time, thereby creating the fundamental standing wave in a circle. You can attempt to create higher harmonics by brushing the pen or pencil at integral multiples of the fundamental frequency. For best results don't leave the pen or pencil tip inserted between the coils, but merely use it to push the coils back and forth.

## J.10. Rubbing the edge of a wine glass

**Demonstration**
An explanation of why a wine glass sings (resonates) when you rub its edge with a moist finger leads to a surprising area of physics: the Bohr model of the atom.

**Equipment**
Three wine glasses. Drinking glasses may also work, but they don't produce as loud a sound as thin-walled wine glasses. Try to find two wine glasses with the same thickness but different diameter, and two with the same diameter but different thickness.

**Comment**
Hold the glass by its stem or base as you rub a moist finger on its rim. You may find that you have to rub for an ex-

tended period of time before you are able to produce a loud sound. The sound waves created by rubbing the edge of a wine glass resonate in the glass itself rather than in the air space inside. A key observation, which shows that the air column is not the source of the resonance, is that the frequency of sound *decreases* as water is added to the glass. A clear sound is heard even when the glass is entirely full of water.

The wave in the glass is excited as your finger alternately slides and sticks while moving along the rim, causing the glass to vibrate in a longitudinal mode (vertical lines on the glass vibrate back and forth). The longitudinal wave creates a resonance in the glass. It may be a misnomer to speak of the wave as a standing wave, because by moving your finger in one direction around the edge you create a wave traveling in one direction. The wave amplitude grows if the wavelength equals the circumference of the glass, the resonance condition for a wave traveling in a circle. Thus for two glasses of identical thickness, the glass with the larger diameter produces sound of longer wavelength and lower frequency. The assumption that the wavelength equals the circumference of the glass is obviously an oversimplification because the wave is not one dimensional, and the glass diameter varies with height. Also, it is possible that higher harmonics are present, in which case an integral number of wavelengths would have to fit around the circumference, just like the boundary condition for the Bohr model of the atom.

If you rub two glasses of the same diameter and height but different thickness, the one with the greater thickness produces sound of lower frequency. The frequency is lower in this case because the thicker glass—having greater mass per unit length—results in a lower sound speed. Therefore the frequency is lower, since the wavelengths are identical for glasses of the same circumference.

How can the decrease in frequency as water is added to a wine glass be explained? More water results in more viscous drag on the glass, i.e., more water molecules dragged along with the glass molecules as they vibrate, or a higher effective mass per unit length. The higher mass per unit length results in a lower frequency, just as in the case of thick-walled glasses. One observable consequence of the viscous drag is that you may see a water wave following your finger as it moves around the edge of the glass. The water wave is most easily seen when the glass is made to sing loudly, and the water surface is viewed up close and edge on.

## J.11. Longitudinal waves in a rod

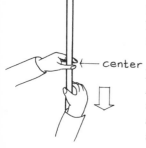

center

**Demonstration**
By stroking an aluminum rod, you can excite longitudinal standing waves, causing the rod to "sing."

**Equipment**
A 6-ft-long, ½-in.-diameter aluminum rod; and some "no-slip" spray or gel (obtainable at sporting-goods stores). The aluminum rods used for lab stands also work well, but being shorter produce higher frequencies.

**Comment**
The longitudinal waves for a thin rod that is free at both ends satisfy the same boundary conditions as the sound waves in an air column open at both ends—namely, the wave displacement must have an antinode at each end. The fundamental mode in both cases has a node at the center, so the length of the rod $L$ corresponds to one half-wavelength (or $n$ half-wavelengths for the $n$th harmonic). The fundamental frequency is the speed of sound in aluminum divided by twice the length of the rod, and the $n$th harmonic frequency is $n$ times that of the fundamental. To excite the fundamental vibration mode and make the rod "sing," hold the rod at its exact center (you might want to make a mark there with a felt-tip pen) and stroke the rod repeatedly with your other hand. You will need to spray or rub the no-slip spray (or gel) on your hand to produce enough friction to cause the rod to vibrate. Repeated stroking should cause an audible sound to be emitted. If you have difficulty, try stroking at different speeds (slow stroking seems to work best).

In principle, the sound emitted should be a mixture of all standing waves that have a node at the middle (the odd harmonics), but in practice the fundamental (first harmonic) is by far the dominant frequency. You can demonstrate this fact after the rod is singing by pinching the rod with two fingers at a point 16⅔ percent from one end, thereby causing this point to be a node. The fundamental, but not the third harmonic, is suppressed, because only the latter has nodes at the 16⅔-percent points. The silence created when the singing rod is pinched at one of these points shows that no third harmonic is produced when the rod is held at the center and stroked. You can, however, directly excite the third harmonic by holding the rod at one of the 16⅔-percent points rather than at the center when you stroke it. In general, you can excite the $n$th harmonic by holding the rod at a point a distance $L/(2n)$ from one end, while stroking the rod.

## J.12. Transverse waves in a rod

**Demonstration**
The fundamental mode of transverse standing waves in a rod can be excited with a karate chop to the middle.

**Equipment**
A 6-ft-long, ½-in.-diameter aluminum rod.

**Comment**
Transverse waves in a rod are more complex than longitudinal waves, since they do not have a simple sinusoidal form, and they satisfy a fourth-order wave equation (longitudinal waves obey a simpler, second-order wave equation). Nevertheless, the wave resembles a sinusoidal shape (see fundamental mode in illustration). The fundamental mode for transverse waves has an antinode at each end but, unlike the longitudinal wave, the center is also an antinode. The nodes occur at points that are 22.4 percent of the length of the bar (16⅛ in.) from each end.

To excite this fundamental mode, hold the rod vertically between two fingers at one nodal point (16⅛ in. from one end) and give it a hard karate chop at the center, causing it to swing away from you. When it swings back, catch it so that it rests horizontally on the edges of your hands, supported at each node. The standing waves should be readily visible. (It would be a good idea to mark the position of the two nodes with felt-tip pen.)

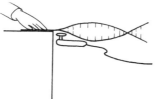

## J.13. Resonant standing waves in a rod

**Demonstration**
Higher harmonics of transverse standing waves in a rod can be resonantly excited using a vibrator.

**Equipment**
A massager/vibrator (see A.4); and a 3-ft-long, ⅛-in.-diameter wooden dowel.

**Comment**

Put the dowel on a table, so that part of it projects out over the edge. Place the vibrator under the rod at a point close to the table, while pressing down firmly on the part of the dowel that is on the table. Press the vibrator tip upward against the dowel. If the length of dowel projecting out over the edge of the table, $L$, is close to the length at which a particular harmonic resonates at the 120-Hz vibrator frequency, that harmonic will be excited to a large amplitude of several inches. Otherwise, one or more harmonics will be excited, and the amplitude will be small. You can excite the various harmonics, beginning with the fundamental, by starting with a small length $L$, and gradually increasing it a few inches at a time, observing what modes are excited in each case. The fundamental has an antinode at the dowel's free end, and the $n$th harmonic has $n$ nodes along the rod, including the node at the table end. The shape of the waves in a clamped rod is not sinusoidal, since they satisfy a fourth-order wave equation, not the usual second-order equation. The lengths $L$ that should exhibit resonance for harmonics 2, 3, 4, . . . should be the following multiples of the fundamental length: $\frac{5}{3}$, $\frac{7}{3}$, $\frac{9}{3}$, . . .

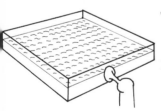

## J.14. Standing waves on the surface of a liquid—Part I

**Demonstration**

Short-wavelength standing waves on a water surface can be generated by pressing the tip of a vibrator against a transparent box full of water.

**Equipment**

A massager/vibrator (see A.4); a clear plastic rectangular storage box (the size is not critical, but 5×5×4 in. would be a good size); and a small rectangular plastic box (with different length and width dimensions) that can fit within the larger box (again, the size is not critical, but 1½×3×4 in. would be good—4 in. being the height).

**Comment**

With the large box filled with water (and placed on an overhead projector for display to a large group), standing surface waves can be excited using the vibrator. For best results try various projector focus settings and various degrees of contact between the vibrator and the side of the box. The wave patterns will be fairly complex because the vibrator pressure is not applied uniformly to the side of one wall to generate plane waves. If the applied pressure is gentle, however, you may see plane-wave patterns.

One way to get excellent plane waves is to fill the small rectangular box with water and place it in the middle of the larger box (also filled with water). Apply the vibrator to the side of the larger box. The applied pressure fluctuations on the wall of the smaller box are *not* localized at one point, and the condition that an integral number of half-wavelengths fit between the walls of the smaller box results in beautiful standing plane waves.

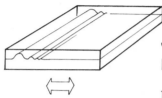

## J.15. Standing waves on the surface of a liquid—Part II

### Demonstration
Long-wavelength standing waves can be created on a water surface by moving a partially filled tray back and forth in simple harmonic motion. Dispersion of wave pulses can also be shown.

### Equipment
A transparent rectangular tray or storage box filled with water to a depth of about ½ in. This shallow depth results in a suitably slow wave velocity.

### Comment
If you repeatedly shake a water-filled tray back and forth at a regular frequency, you cause a series of reflected wave pulses that approximate standing waves. The resulting wave pattern can easily be seen if the tray is on an overhead projector. The easiest standing-wave mode to excite is the fundamental, for which you must time your oscillations of the tray to match the round-trip time of a reflected wave. You may also be able to excite higher harmonics by shaking the tray at a small integral multiple of this frequency. If the tray is rectangular rather than square, you can excite waves traveling along either the long or short dimensions. You can also demonstrate dispersion of traveling-wave pulses using the

water-filled tray. Wave pulses can be described mathematically as a superposition of waves of different frequencies. Water pulses disperse because the various frequency components making up the pulse travel at different speeds, causing the pulse to broaden gradually as it travels. To observe this phenomenon, quickly move one edge of the tray up and down once (by about an inch).

## J.16. Standing waves on the surface of a liquid—Part III

### Demonstration
Circular standing waves can be created in a water-filled pie dish by dunking a pencil at the right frequency.

### Equipment
A transparent pie dish about 9 in. in diameter, filled with water to a depth of ½ in.; a pencil; and a metronome, if available.

### Comment
Place the water-filled pie dish on an overhead projector and dunk the eraser end of a pencil, creating an expanding circular wave pulse. When the expanding wave reflects off the walls of the pie dish, a reflected contracting wave pulse is created. If you dunk the pencil at the center of the dish, the reflected wave contracts back to the center. If the pencil is dunked at a point that is slightly off center, the reflected wave contracts back to a symmetrically located point on the other side of the center. Dunk the pencil at a regular frequency so that the oscillation period just matches the round-trip time for a wave. (Put the pencil back in the water just as the contracting wave returns to the center.) This is the fundamental mode for circular standing waves. A metronome may help you maintain a regular frequency of pencil dunking. A metronome would also be useful in trying to excite higher-harmonic standing waves, whose frequencies are integral multiples of the fundamental frequency. For best results, use a water depth of about ½ in., and try varying the projector focus off the water. Strictly speaking, the waves produced in this demonstration are not quite standing waves, because the dunking pencil produces waves that are pulsed rather than sinusoidal in shape.

# J.17. Standing waves on a soap bubble

### Demonstration
Standing waves on a membrane can easily be excited to large amplitude by shaking a frame after immersing it in a soap-bubble solution.

### Equipment
Two plastic straws; some cotton string; a 9×11-in. pan 1 to 2 in. deep; and soap-bubble solution (available from Toys-R-Us for under $1.00). You can also make your own bubble solution using 185 parts pure glycerine (obtainable from a drugstore), 70 parts Dawn or Joy detergent, and 225 parts water. You should use distilled water if your tap water is hard.

### Construction
Use the two straws and the string to make a frame to hold the bubble film. Thread the string through each straw and tie a knot to make a rectangular frame about 9×12 in. Tuck the knot inside one of the straws to get it out of the way. Dip the frame into the bubble solution in the tray and be sure that everything—including your fingers—is coated with bubble solution. Gently lift the frame out of the solution with the straws together, holding one straw in each hand. Slowly pull the straws apart. The bubble film should fill up the entire frame. Be sure to keep the straws straight; don't pull them too hard or they will flex outward.

### Comment
Exciting the fundamental standing-wave mode (which has a maximum in the center) is easy. Simply start shaking the frame up and down gently, and the oscillations will grow to appreciable size if you are shaking at the resonant frequency. You may be able to excite other modes by rocking the frame back and forth in a seesaw motion. If you are showing this demonstration to a large group, the visibility of the soap bubble film can be increased by shining light on it; this will cause colors—due to interference—to appear in the thin bubble film.

# Doppler Effect and Beats

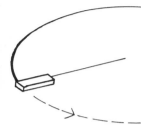

## K.1. Whirling a beeper in a circle

**Demonstration**
A beeper twirled at the end of a string in a horizontal circle
causes a Doppler-shifted sound that rises and falls in frequency during each rotation.

**Equipment**
A beeper of the kind described in demonstration J.5.

**Comment**
The wire that connects the beeper to the battery can serve as
the string you use to whirl the beeper in a circle. Be sure the
wire is firmly attached, however, since you whirl the beeper
in the direction of the listeners. As a precaution you should
make a loop at the end of the wire through which to stick
your hand. One final precaution: Be sure, when you are
ready to whirl the beeper at the end of the wire, that you are
well away from any walls, people, or other objects that the
beeper might strike.

The fractional increase or decrease in frequency during
those parts of the circle when the beeper is moving directly
toward or away from the listeners is given by $\pm v/v_s$, where $v$
is the beeper speed and $v_s$ is the speed of sound. This
amounts to a fractional change in frequency of $\pm 10$ percent
or $\pm 20$ percent for a beeper whirled at the end of a 1-m-long
wire at 5 or at 10 revolutions per second, respectively.
Clearly, the magnitude of the effect (and the risk of hitting
someone with a slingshot projectile!) increases with the
speed at which you whirl the beeper. As the beeper rotates,
listeners should hear a continuously fluctuating frequency,
varying between the above limits every $T$ seconds, where $T$
is the rotation period. Listeners should be told to try to detect
the change in pitch, not the change in loudness that also
occurs.

This demonstration can serve as a model for the method
astronomers use to search for planets around stars other than
the sun. If such a planet existed, it and its companion star

would orbit a common center of mass. Let us suppose the plane of rotation is viewed edge-on from earth. At one point during the star's rotation about the center of mass, it has an extra velocity toward the earth, while a half-revolution later its velocity direction is reversed. When the star moves toward the earth its light spectrum is blueshifted (shifted toward higher frequencies), and when it moves away its spectrum is redshifted. Even though the (nonluminous) planet cannot be observed directly, its presence can be ascertained through the Doppler shift it causes in the light from its companion star. Given current detector sensitivities, this technique can be used only to detect quite massive planets, and it would fail to detect a planet with a mass comparable to that of the earth.

## K.2. Shaking a rod vibrating in its longitudinal mode

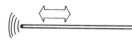

### Demonstration
A "singing rod" (see demonstration J.11), pointed toward a listener and rapidly moved back and forth along the direction in which the rod is pointing, creates a Doppler-shifted sound that rises and falls in frequency each oscillation.

### Equipment
A 6-ft-long, ½-in.-diameter aluminum rod; and some "no-slip" spray or gel (obtainable at a sporting-goods store).

### Comment
Cause the rod to "sing" (as explained in J.11). Holding the rod at its center like a javelin, rapidly move it back and forth, with one end facing the listener. The Doppler effect will give rise to a fractional change in frequency of $\pm v/v_s$, where $v$ is the maximum speed of the rod and $v_s$ is the speed of sound. Unlike demonstration K.1, the complicating effect of a periodic change in loudness accompanying a change in frequency is not present in this demonstration.

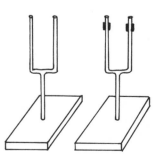

## K.3. Beats using two sound sources

### Demonstration
Beats are produced from two sound sources if they are close in frequency.

**Doppler Effect
and Beats**

**Equipment**

Two sound sources that have frequencies that differ by no more than 20 Hz. For frequency differences greater than about 20 Hz, there are too many beats per second to hear the beats distinctly. One possibility would be to use two identical tuning forks, and change the frequency of one slightly by sticking a bit of masking tape on it to lower its frequency. The kind of tuning forks mounted on resonating boxes work best, since they produce a loud, long-lasting sound, and they need not be hand held. These box-mounted tuning forks often come with a sliding piece of metal on one fork that you can use instead of tape to alter the frequency.

**Comment**

The beat frequency, which is equal to the difference between the two source frequencies, can be varied by changing the amount or placement of tape you use. (Be sure to use masking tape, not transparent tape, which is too hard to remove from the fork.) You should put tape on each of the two tines of one of the forks, at the same height. The higher on the fork you put the tape, the greater the resulting reduction in frequency.

If the tuning forks are completely identical, you should observe no beats when neither fork is taped. You should also find that the closer you place the tape to the top of one of the forks, the higher the beat frequency is.

If the tuning forks are not identical, adding tape to one of them may give rise to either more or fewer beats per second, depending on whether the tape was added to the lower- or higher-frequency fork—something you will probably not know at the outset.

If you find using tuning forks somewhat inconvenient (since their sound fades away rather quickly), you may wish to try using two beepers as your sources (the kind used in demonstration J.5). The likelihood is, however, that the two beeper frequencies will not be sufficiently close for you to be able to hear distinct beats, since they are only standardized to within ±14 percent.

# Electricity

## L.1. Solar-powered fan

**Demonstration**

A small fan can be powered by a solar collector, illustrating two energy conversion processes.

**Equipment**

Either a "solar energy project kit" (sold, for example, from Radio Shack for about $10.00) or a solar collector hat (which can also be purchased for around $10.00).

**Comment**

The solar-collector-powered fan can be shown to a large group by placing it on an overhead projector and making the fan blades visible. By sliding a card underneath the solar collector, the fan can be started and stopped, showing that the current produced depends on the area of solar collector that is exposed to light.

## L.2. Lemon battery

**Demonstration**

You can make a battery from two electrodes and a lemon, measure its electromotive force (EMF) and internal resistance, and use it to activate a buzzer.

**Equipment**

A low-voltage buzzer; a lemon (and several other fruits); a multimeter; and copper and zinc electrodes, cut to dimensions that allow them to be inserted into the lemon. A suitable buzzer would be the model SBM2 8715 Sonalert made by Mallory, and sold by Radio Shack, for example.

**Comment**

To make a lemon battery, insert the copper and zinc electrodes into the lemon about 90° apart. The voltage of the lemon battery can be expressed in terms of its EMF $E$ and internal resistance $r$ as $V = E - ir$, showing that the voltage depends on the current drawn, $i$. If you measure the voltage across the electrodes with a multimeter, you measure the EMF $E$, since the current drawn is negligible. If you switch the multimeter to a current scale and (momentarily) connect it across the battery, you will short it out. In this case the current reading will be $i_{max} = E/r$, because the voltage $V$ across the battery terminals is zero. The battery's internal resistance can then be found from $r = E/i_{max}$. You should find that the lemon battery has an EMF of a bit under one volt and an internal resistance of several thousand ohms, but—as with any battery—if you short it out for any length of time, you run it down, causing $i_{max}$ to decrease and $r$ to increase.

Prepare a fresh lemon battery, and connect it to a low-voltage buzzer, which should produce an audible sound. You may wish to prepare batteries using other fruits in place of the lemon to see what they sound like.

## L.3. Human battery

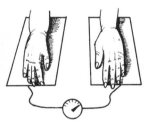

**Demonstration**
A battery that consists of a pair of electrodes and a person (the electrolyte) produces a measurable voltage, but only if the electrodes are made of different metals.

**Equipment**
Two sheets of copper and one sheet of zinc (actually any pair of different metals will do); and a DC voltmeter.

**Comment**
Connect the terminals of the voltmeter to the copper and zinc sheets, and put one hand on each sheet. The voltmeter should read about 0.7 V. Other pairs of metals will yield different values, depending on the metals' relative electro-chemical potential. If you replace the zinc sheet by the second copper sheet, and place each hand on a copper sheet, the voltmeter will of course read zero, since two different metals are required for a battery. In order for the battery to function, the surface of your hands must have plenty of ions (this is generally the case). If copper and zinc strips cut from the sheets are placed in a beaker or jar of distilled water, the voltage reading will be zero, showing that an electrolyte (ions in solution) is necessary, as well as two different metals, for a battery to work.

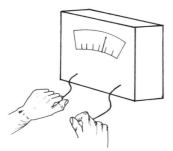

## L.4. Human resistance

**Demonstration**
You can measure your own resistance by holding the probes of a multimeter.

**Equipment**
A multimeter. (Inexpensive ones can be obtained from hardware stores for under $15.00.)

**Comment**
The nature of the contact between one's fingers and the probes is the primary factor in determining the measured resistance. I obtained readings anywhere from 0.2 to 1.0 million ohms, depending on how hard I squeezed the probes. Moreover, the readings didn't change much whether I held one probe in each hand or touched the two probes an inch apart on one finger. Moistening my fingers did, however, make a significant difference in the resistance, lowering it to 80,000 ohms. So-called "lie detectors" rely in part on the measured resistance of the skin, which varies according to perspiration, supposedly an indicator of lying.

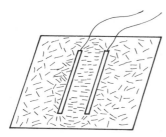

## L.5. Mapping the electric field

**Demonstration**
The electric field that surrounds a variety of charged conductors can be mapped using grass seed.

**Equipment**
A high-voltage, low-current DC source; grass seed; two 100-g weights; two metal rods of square cross section; and a "tin" can (tuna-fish-can size, with top and bottom removed). Van de Graaff generators and Wimshurst machines are good high-voltage, low-current sources.

**Comment**
You can create an electric field pattern on an overhead projector with a high-voltage source connected by wire to a pair of conductors placed on the overhead projector. Some good patterns can be obtained using two closely spaced parallel metal rods (to approximate two parallel plates), two 100-g weights (to approximate two point charges), and one weight placed at the center of the tuna-fish can (to approximate concentric cylinders). With one conductor connected to the

high-voltage source and the other connected to ground (the overhead projector), sprinkle grass seed all over the top of the projector. If the voltage of the source is sufficiently high, the electric field will orient the seed, making the field lines visible. The field pattern appears more clearly when you use a lot of grass seed, and sprinkle it *after* the high voltage is turned on. To avoid making a mess you may wish to put the conductors and grass seed on a piece of transparency film rather than directly on the overhead projector.

## L.6. Charge-propelled aluminum can

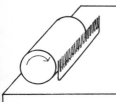

### Demonstration
An empty aluminum soda can rolls toward a charged object because of the electric charges induced on the can when the object is brought near.

### Equipment
An aluminum soda can and a pocket comb. You may find that a charged comb does not supply enough force to make the can roll, in which case you could either (a) use other more conventional means of putting a somewhat greater charge on an object, e.g., rubbing a rubber rod with fur, or (b) use a light-weight hoop, which rolls more easily than the can. To make a 2-in.-wide aluminum hoop, stab the can with a sharp knife, and cut it all the way around with a pair of scissors.

### Comment
Charge the comb by running it through your hair and place it, with its teeth vertical, next to the can on an overhead projector (or any horizontal surface). The comb attracts the uncharged can because of the charges induced on the can. The can continues to roll as long as the comb is kept just ahead of the can without making contact. The comb needs to be held with its teeth vertical, so that a maximum amount of the comb's surface area is next to the can. However, the charge on the comb may not be great enough in humid weather, so you should also have a hoop ready just in case.

## L.7. Light bulbs in series and in parallel

### Demonstration
Predicting the relative brightness of three light bulbs can pose a challenge to one's understanding of circuits.

## Section L

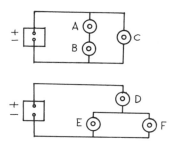

### Equipment
A six-volt lantern battery; three miniature ceramic sockets; three miniature light bulbs (six-volt size); and some stiff wire. All of these items are obtainable at a hardware store.

### Comment
The miniature sockets are small enough that the circuits shown in the illustration can be wired on an overhead projector, allowing all the wiring connections and the bulb sockets to be seen easily in silhouette. If stiff wire is used to make the circuit connections, it can be bent in straight-line segments to resemble a circuit diagram on the overhead projector. The difference between lit and unlit bulbs can easily be seen in the projected image, but the bulbs' relative brightness can be more clearly seen when the bulbs are viewed directly.

Before making the final connection to the battery, give people a chance to predict the relative brightness of the three bulbs (A and B should be equally bright, and less bright than C.) Before you rewire the bulbs to make the second circuit, give people a chance to predict how unscrewing bulb A affects the brightness of the other two bulbs (B goes out and C is unchanged). After wiring the second circuit, ask for predictions of the relative brightness of D, E, and F before making the final battery connection (E and F are both much fainter than D). Finally, ask how unscrewing bulb F affects the brightness of D and E (D decreases and E increases). In each case a correct prediction can be made by computing the current through the bulbs, assuming that they all have the same resistance. Nevertheless, it is surprising how often people who have an excellent facility with circuit problems fail to make the correct predictions.

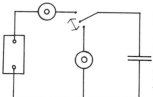

## L.8. Discharging a capacitor through a light bulb

### Demonstration
When a very large capacitance (1 F) is charged or discharged through a light bulb, the current and the time constant are large enough that the exponential variation of the current is evident from the brightness of the bulb.

### Equipment
A one-farad, five-volt capacitor (available from Nakamura, Inc. for about $11.00); a six-volt lantern battery; two min-

iature light bulbs; two miniature ceramic sockets; and stiff wire. All of these items (except for the capacitor) are available in hardware stores. Don't forget to discharge the capacitor before putting it away. It is rated at five volts, and it can be charged to six volts only for short periods of time.

**Comment**

The components used in this demonstration are all small enough that the circuit in the illustration can be wired on an overhead projector with all wiring connections and the brightness of each bulb clearly seen. Note that initially one wire to the capacitor is not connected to either the upper or lower light bulb. When this wire is connected to the terminal on the upper socket, that bulb begins to glow brightly, and then it gradually fades as the capacitor charges. After the capacitor is charged, if you switch the top wire on the capacitor over to the lower socket, that bulb begins to glow brightly, and then gradually fades as the capacitor discharges. In both cases, the time variation of the current should follow the exponential decay law. You don't see a true exponential decay when you view the bulbs, because their apparent brightness is not proportional to the current, but you do get a feeling for the exponential decay from the gradual drop in brightness with no sharp cutoff.

The rate of exponential decay is determined by the time constant $RC$, which is the time it takes for the current to reach $1/e$ (about 37 percent) of the initial value. You can see the effect of doubling or halving the time constant by allowing the capacitor to charge or discharge through two bulbs in series or two bulbs in parallel. A change in time constant must be accompanied by a corresponding change in bulb brightness: if bulbs burn longer, they don't burn as brightly, because the capacitor stores a fixed energy (eighteen joules) when charged up to six volts, whether it discharges slowly through a large resistance or rapidly through a small resistance.

## L.9. Discharging a capacitor through a voltmeter

**Demonstration**

By allowing a capacitor to discharge through a voltmeter you can measure the time constant and show the exponential time dependence of the discharge.

**Equipment**

A capacitor; a multimeter or voltmeter; a 12 V battery; and a stopwatch. You should use a capacitance such that the time constant $RC$ is somewhere in the range of 10 to 60 seconds. $R$ is the voltmeter resistance on the 10 V scale, which can be found either in the meter manual or measured with a multimeter. It would be better not to use a digital voltmeter because seeing the rate at which a needle swings can give a better sense of the exponential decay than watching numbers change. A transparent projection meter would allow the demonstration to be given on an overhead projector.

**Comment**

Charge the capacitor up to 12 V by connecting it to the battery, and then connect it to the voltmeter set on the 10 V scale. Start your clock just when the reading begins to go below 10 V. Record the times $T_1$, $T_2$, $T_3$, and $T_4$, when the voltage reaches 5.0 V, 2.5 V, 1.25 V, and 0.625 V, respectively. If the discharge is exponential, you should find that the four recorded times represent one, two, three, and four half-lives, where the decay half-life is given by 0.693 times the time constant $RC$.

## L.10. Conversion of electrical to thermal energy

**Demonstration**

A comparison of electrical and thermal energies can be made by measuring the temperature rise that an immersion heater causes in a known quantity of water.

**Equipment**

An immersion heater; a thermometer (preferably the digital type with an external probe); and a Styrofoam cup.

**Comment**

Submerge the heater in a cup containing a mass $m$ of water, and leave it plugged in for a time $t$ while the thermometer records the temperature rise $dT$. The thermal energy (in joules) gained by the water is given by $4.186m(dT)$, with $m$ in grams and $dT$ in degrees Celsius. You can compare the thermal energy with $Pt$, the electrical energy consumed, where $P$ is the heater's rated power. If you don't know the rated power, simply measure the heater's resistance with a multimeter, and compute the power from $110^2/R$.

## L.11. Superconductivity

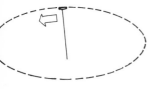

**Demonstration**
Magnetic levitation of a small magnet above a superconducting disk demonstrates the Meissner effect.

**Equipment**
A superconductivity demonstration kit (Edmund Scientific Company sells one for about $25, for example); and some liquid nitrogen. When working with the liquid nitrogen be sure to avoid splashing it, and do not touch any items that were immersed in it until they have warmed up. Liquid nitrogen, which is as cheap as milk, is difficult to obtain in amounts less than thirty liters from bottled-gas companies, but you may be able to obtain a small quantity from a local university chemistry or physics department, a hospital, or a dermatologist.

**Comment**
Pour a small amount of liquid nitrogen into a dish or Styrofoam cup until it is about a quarter inch deep, and wait for the boiling action to stop. Using nonmetallic tweezers, place the superconducting disk in the liquid so that its top is flush with the liquid surface, and wait for the renewed boiling to stop. Using the tweezers, place the small magnet that comes with the kit at the center of the superconducting disk. Instead of settling down on the disk's upper surface, the magnet will float above it—a demonstration of the Meissner effect, which is associated with superconductivity. The Meissner effect occurs because eddy currents induced on the surface of the superconductor exactly cancel the magnetic field produced by the magnet in the interior of the superconductor. This magnetic field cancellation results in essentially an exact mirror-image magnet below the superconductor, which accounts for the levitation.

## L.12. Twirling a neon lamp on a line cord

**Demonstration**
The alternating nature of current from a wall outlet can be shown by the segmented visual image created when a

neon lamp is rapidly twirled in a circle at the end of a line cord.

**Equipment**
A small neon lamp, with a resistor already soldered to one lead (obtainable from Jerryco, Inc. for about $2.00); and a line cord with plug.

**Construction**
The leads of the neon lamp and resistor need to be soldered to the wires of the line cord and carefully wrapped with tape or shrink-wrap to avoid a short circuit.

**Comment**
When the lamp is twirled in a circle, the number of dashes you see depends on the rotational speed. The lamp is illuminated during a part of both the positive and negative portions of the 60-Hz AC voltage cycle, as long as the voltage is high enough, so it flickers at 120 Hz. If the lamp is twirled at 5 revolutions per second, for example, 24 lit segments are seen around the circumference of the circle. You can't see the lamp flickering when it is stationary, because images that occur at a rate faster than about 16 per second cannot be resolved by the eye as separate. If you tried a similar demonstration with an ordinary incandescent bulb twirled in a circle, no segments would be seen, because incandescent bulb filaments don't have a chance to cool off between the 120-Hz voltage peaks.

## L.13. Departures from the inverse-square law

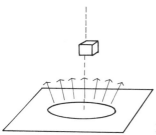

**Demonstration**
It is only for a point charge that the electric field has an inverse-square dependence on distance. Departures from the inverse-square law for the electric field of a charge of finite size can be simulated using a luminous disk.

**Equipment**
A meterstick; a light meter; and a large piece of opaque cardboard with an 8-in.-diameter hole over which a translucent piece of paper is taped. When the cardboard is placed on the overhead projector, the result is a reasonably uniform disk-shaped light source.

**Comment**

The electric field of a point charge varies as the inverse square of the distance from the charge. Likewise, the light intensity from a point source of light also has an inverse-square dependence on distance. The departures from the inverse-square law due to finite source size are not exactly the same in both cases, however, owing to the vector nature of the electric field and the scalar nature of the light intensity. For example, the electric field at the center of a ring of charge must be zero because of the vector cancellation of contributions from all around the ring, but the light intensity at the center of a luminous ring is obviously not zero, because no such cancellation occurs. Despite the important differences between the two cases, the easily observed departure from the inverse-square law for a luminous disk aids in understanding the similar departure for a disk of charge.

To observe the effect, place the cardboard with cutout hole and translucent paper on the overhead projector, being careful that no light escapes at the edges of the cardboard. Position the light meter directly above the center of the hole, and measure the light intensity $I$ at two-inch intervals above the hole, from a height of zero to a height of 24 inches. Multiply each of the intensity readings by the square of the corresponding height $r$ above the hole. Plot the thirteen values of the product $Ir^2$ against $r$. These values should lie on a horizontal line only when the source can be considered a point, which will obviously not be the case for distances that are small compared to the hole's radius.

# Magnetism

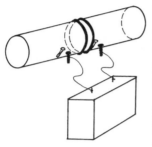

## M.1. Force between two coils

### Demonstration
The magnetic force between two current-carrying conductors can be shown using two adjacent coils free to slide on a transparent cylinder.

### Equipment
Magnet wire (0.16 ohms/ft); a rolled-up transparency blank; and a heavy duty 12 V lantern battery.

### Construction
Wind two 2-in.-diameter coils from magnet wire. On the last of 50 turns wrap the wire around the coil to hold all the turns in place, leaving two long leads. Roll up a transparency blank to make a cylinder on which you can place the two coils and have them slide very easily. Using four screws with nuts, as shown, construct four legs on which the cylinder can rest. Connect the coil wires to two of the legs so that

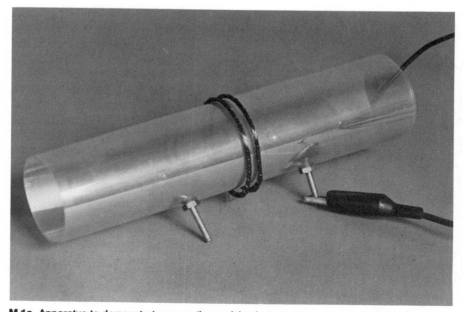

**M.1a** Apparatus to demonstrate magnetic repulsion between two current-carrying coils the two coils are in parallel. The currents in the two coils should go in opposite directions. Use double nuts on each leg to make good electrical contact with the leads. Also, the ends of the magnet wire must be scraped well.

**Comment**
Place the two coils as close together as possible and momentarily make battery contact with the two screw legs. This should cause the coils to jump apart a short distance. The transparent demonstration is easily shown on an overhead projector. It is not essential that the two coils be placed on the plastic cylinder; they could each be suspended from a pair of wires like two trapezes, and held next to each other. However, the cylinder arrangement is less fragile, and easier to store and set up.

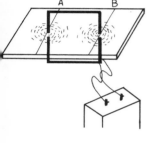

## M.2. Field of a long straight wire

**Demonstration**
The magnetic field of a segment of a straight wire can be mapped using iron filings and compasses.

**Equipment**
A piece of acrylic sheet (obtainable from a hardware store or a plastics company); magnet wire (0.16 ohms/ft); iron filings; transparent compasses; a heavy duty 12 V lantern battery; and plastic cement.

**Construction/Comment**
The vertical rectangular loop shown in the illustration is a tightly wound coil rather than a single loop of wire. The demonstration could be done using just a single loop, but only if a high-current source is used. In order to achieve the same magnetic field strength using a single loop of wire, the source would have to supply 50 times the current necessary for the 50-turn coil. The coil used here is rectangular, with two vertical segments that pass through holes drilled in an 8½×11-in. acrylic sheet. When a current is passed through the coil, each vertical segment produces circular magnetic field lines in the horizontal plane of the acrylic sheet. These circular field lines can be mapped using iron filings or compasses. Transparent compasses work well on an overhead projector. The two horizontal segments of the rectangular coil also produce a magnetic field in the plane of the sheet, so the field lines are circular only in the immediate vicinity of the two vertical coil segments.

158

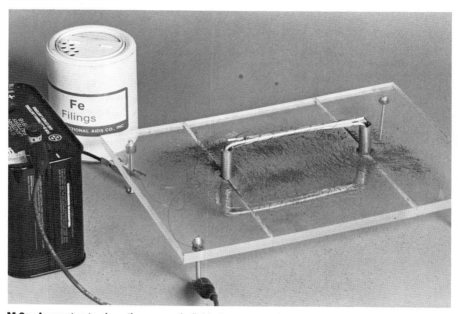

**M.2a** Apparatus to show the magnetic field of a current-carrying rectangular coil

Follow these steps to make the coil:

(1) Cut an 8½×11-in. rectangular piece out of the acrylic sheet.

(2) Drill two ¼-in.-diameter holes on a center line three inches from each 8½-in. edge and separated from each other by five inches.

(3) Cut the acrylic sheet along lines *A* and *B*, which pass through the holes (see illustration). This will make it a lot easier to wind the coil that passes through the two ¼-in. holes.

(4) Wind 50 turns of magnet wire into a 2×5-in. rectangular coil, leaving long leads.

(5) Make a frame the size and shape of the rectangular coil out of stiff wire, such as that from a coat hanger.

(6) To make the coil structure rigid, wrap one of the long leads from the coil around both the rectangular coil and the wire frame.

(7) With the vertical sides of the coil placed through the holes, glue the three pieces of acrylic sheet back together with plastic cement.

(8) Put mounting screws in holes in the four corners of the acrylic sheet, so that the bottom of the rectangular coil will not touch the overhead projector on which it is placed.

**(9)** Connect the two leads from the coil to two of the screws. Use two nuts on each screw to make a good electrical contact with the coil leads.

**(10)** Sprinkle iron filings on the acrylic sheet and connect a 12 V lantern battery to the coil leads via the screws. Tap the acrylic sheet with a hard object to reveal circular magnetic field lines around each vertical coil segment. Don't leave the battery connected longer than is necessary to create the pattern.

**(11)** Transparent compasses placed around each vertical coil segment also show the directions of tangents to the field lines at each point.

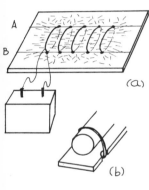

(a)

(b)

## M.3. Field of a coil

**Demonstration**
The magnetic field of a coil can be mapped using iron filings and compasses.

**Equipment**
Same as M.2.

**Construction/Comment**
Unlike a standard helical coil, the coil used here is bunched in five circular coils of 50 turns each, connected by one continuous wire. The idea is to have plenty of unobstructed open space between the five segments so that the iron filings can easily be seen when the coil is placed on an overhead projector. The construction method is similar to that used in demonstration M.2:

**(1)** Cut an 8½×11-in. rectangular piece from an acrylic sheet.

**(2)** Drill two rows of five ¼-in.-diameter holes one inch apart; the two rows should be two inches apart and positioned on the sheet so that the wound coil, which will pass through the holes, is centered on the sheet.

**(3)** Cut the acrylic sheet along lines $A$ and $B$, which pass through the two rows of holes. (This will make it a lot easier to wind the coil.)

**(4)** Wind the coil in five segments of 50 turns each, using the magnet wire. Each segment should be as circular as you can make it. One way to ensure that each of the 50 turns has the same radius is to place two one-inch-diameter cylinders above and below the two-inch-wide piece of acrylic on which you are winding the coil, and wind the coil around

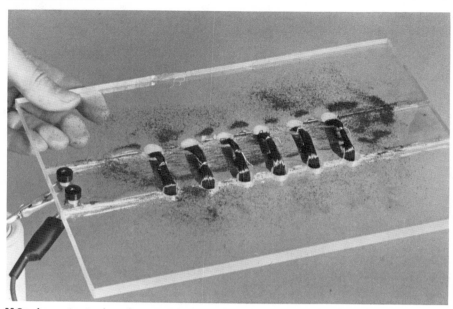

**M.3a** Apparatus to show the magnetic field of a five-segment current-carrying solenoid

these cylinders [see part (b) in the illustration]. Pull the cylinders out each time you complete a 50-turn segment. Remember that all five segments are made from one continuous piece of wire.

(5) After all five segments have been made, wrap additional wire around each segment to hold it together tightly.

(6) Follow steps (7) through (11) of demonstration M.2.

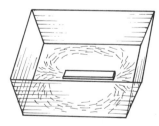

## M.4. Field of a magnet

### Demonstration
Iron filings can be used to map out the magnetic field of various permanent magnets.

### Equipment
Magnets; iron filings; and a deep transparent plastic storage box or dish—the kind used as a cake or salad cover.

### Comment
The plastic dish is important because you can sprinkle iron filings on the dish and put the magnets underneath it, and thus you avoid getting any filings on the magnets (filings can be extremely difficult to remove from a strong magnet).

If you use a strong magnet, you will also need a lot of filings to obtain a nice pattern. To keep your magnets strong, use "keepers" on them. Try using both bar magnets and ring magnets.

## M.5. Deflection of an electron beam

### Demonstration
The electron beam of a cathode-ray tube is deflected when a magnet is brought near, as required by the $q\vec{v} \times \vec{B}$ formula.

### Equipment
A cathode-ray tube and a magnet.

### Comment
When you bring a magnet near the face of the cathode-ray tube or oscilloscope, the electrons are deflected in a direction perpendicular to both their velocity and the magnetic field, i.e., consistent with the $q\vec{v} \times \vec{B}$ formula. Reversing the magnet polarity reverses the direction of the deflection. Bringing the magnet near the origin of the electron beam at the back of the cathode-ray tube increases the deflection, because the earlier a given force is applied to a moving particle, the greater the change in the particle's final momentum.

## M.6. Deflection of a light-bulb filament

### Demonstration
The long, flexible filament of a special kind of light bulb can be made to oscillate when a magnet is brought near, showing that the direction of the magnetic force depends on the directions of the current and the magnetic field.

### Equipment
A "flicker bulb" and small magnet, available from Jerryco, Inc. for about $2.00. The bulb can be screwed into a standard light socket.

### Comment
According to the formula $F = i\vec{L} \times \vec{B}$, the filament of a bulb powered by a 60-cycle current experiences a magnetic force that reverses its direction 120 times a second. You can easily show how the magnetic force depends on the relative orientation of the $\vec{L}$ and $\vec{B}$ vectors by changing the position and

orientation of the magnet. You can also easily show how the magnitude of $\vec{B}$ affects the force, by moving the magnet to various distances, or using a stronger magnet.

## M.7. The world's simplest motor

### Demonstration
The world's simplest motor can be constructed in less than five minutes.

### Equipment
A "D"-size 1.5 V battery; a small disk-shaped magnet; some wire; and a thick rubber band.

### Construction
You need to make a small field coil (the rotating part of the motor) and two supports in which to place the ends of the field coil. Make the field coil of the magnet by winding 10 turns of varnish-coated noninsulated number-22 wire. Make the ends of the wires that extend from the field coil into hook shapes, as shown in the illustration. Scrape the varnish off the *top half* of the two wire ends. Shape two paper clips (or some stiff noninsulated wire) into two rigid supports that have small loops at the top. With a strong rubber band, hold the two supports fixed against the two ends of the battery. Insert the hook-shaped wires coming out of the field coil into the loops at the tops of the two rigid supports, so that the field coil lies just above the disk-shaped magnet, which is placed atop the battery at its middle. Current flows through the field coil as long as its ends are in electrical contact with the supports.

### Comment
If you give the field coil a little push, it should keep spinning for a while. The motor doesn't require split rings or a commutator because the field-coil wires make electrical contact with the loops during only half of the cycle, and the coil's rotational inertia carries it through the other half of the cycle. The motor can be held sideways on an overhead projector and shown to a large group. Take the field coil off when you're done so you don't run the battery down completely, since the field-coil resistance is very small. For more details on this ingenious demonstration, see the following two articles in *The Physics Teacher* (*TPT*): Rudy Keil, *TPT 17*, 308 (1985), and Scott Welby, *TPT 23*, 172 (1985).

There are other devices that can also lay claim to being the world's simplest motor. Among them is "Top Secret," made by Andrews Manufacturing Company, Inc. of Eugene, Oregon, and sold in novelty stores. This intriguing device consists of a small top that can be given a spin on a platform that comes with the top. An initial spin causes the top to spin for days. The device is actually a motor. It has a battery-powered transistor and coil in its base that provide an alternating magnetic field, and there is a permanent magnet inside the top. The spinning magnet top induces a small current in the coil (see section N), which the transistor amplifies. The magnetic field produced by the amplified current attracts (and accelerates) the spinning magnet top, thereby offsetting frictional losses.

# Induced EMF and Lenz's Law

### N.1. Bringing a coil with a bulb near an AC-powered coil

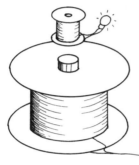

**Demonstration**

A coil powered by 110 V AC induces a current in a nearby coil sufficient to light a small light bulb.

**Equipment**

Two spools of number-22 wire, with lengths of 300 feet and a 100 feet, respectively; a small low-current light bulb; a line cord; and some alligator clips.

**Construction**

If the large (300-foot) coil is not wound on a steel spool, insert some steel rods in its center to increase its inductance. Solder the line cord to the ends of the large coil and wrap the bare wires carefully with tape or shrink wrap. Connect the low-power bulb to the two ends of the small coil using a mini-socket or alligator clips.

**Comment**

When the large coil is plugged in and the small coil is brought near its top, an induced current in the small coil will light the bulb. To avoid overheating the large coil, test its temperature frequently with one finger, and *do not leave it plugged in if it becomes too hot to touch.*

In addition to demonstrating induced EMF's, the AC-powered coil can also be used to show the nature of the magnetic force. If an iron or steel object is brought into direct contact with the steel rods or spool of the large coil, a loud vibration will be heard owing to the 60-cycle variation in current and magnetic field. You can also hear a much quieter 60-cycle noise when the large coil is simply plugged in, because of the magnetic forces between windings of the coil. A similar noise is also heard near large power transformers. If you bring a small disk-shaped magnet near the large coil when it is plugged in, you can feel vibrations in the magnet as it is moved about above the coil.

## N.2. Moving a magnet near a coil with a galvanometer

**Demonstration**
A bar magnet inserted in a coil connected to a galvanometer induces a current in the coil.

**Equipment**
A small coil of 100 turns of wire; a bar magnet; and a galvanometer (the transparent type would be best if you wish to show this to a large group using an overhead projector).

**Comment**
The current induced in the coil when you insert or withdraw a bar magnet is much too small to light a bulb, but should be of sufficient magnitude to register on a galvanometer. The sign of the induced current depends on which end of the magnet you use and whether it is being inserted or withdrawn from the coil. The magnitude of the induced current depends on the speed of the magnet, and is zero for a magnet at rest relative to the coil.

You might also want to try rotating the coil in the presence of a stationary magnet. Rotating the plane of the coil (which changes the magnetic flux through it) produces an induced current, but rotating the coil in a fixed plane does not.

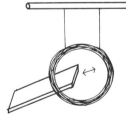

## N.3. Pumping a hanging coil with a magnet

**Demonstration**
A bar magnet moved at the proper frequency in and out of a hanging coil of wire starts the coil swinging, showing that the direction of the induced EMF is given by Lenz's law.

**Equipment**
A coil of wire, and a strong bar magnet.

**Construction/Comment**
Wind the coil using about 50 turns of single-strand wire. Leave 1- to 2-foot leads on each end of the coil, which you can use to suspend the coil from a horizontal bar. Wrap additional wire around the coil to hold it together tightly. Tie the free ends of the coil onto a horizontal bar and connect them electrically with alligator clips or a plain wire. (Attach the

horizontal bar to a rigid support instead of holding it by hand, to avoid the suspicion that your supporting hand is making the coil swing.) "Pump" the coil by inserting and withdrawing a strong bar magnet in a rhythm that matches the natural frequency the coil would have if it were set swinging. You induce an EMF in the coil on both the insertion and withdrawal strokes, so both strokes have to be made with the proper timing, just as you must pump a swing at the right time to increase its amplitude.

Once the coil begins to swing after some repeated pumping, you should insert the magnet as the coil swings away, and withdraw the magnet as the coil swings toward you. It will be easiest to set up the right rhythm if you visualize *pushing* the coil when you insert the magnet and *pulling* the coil when you withdraw it. According to Faraday's law for induced EMF's, the amount of the push or pull will vary with the speed of insertion and withdrawal. This demonstration verifies that the sign of the induced EMF obeys Lenz's law, i.e., that the induced EMF always opposes the change in magnetic flux by making the coil into an electromagnet that attracts or repels the bar magnet so as to always oppose its motion. You can also remove the alligator clips that make the coil a closed loop; no effect is observed in this case because only a momentary current flow through the coil can occur without a complete circuit.

When performing this demonstration before a large group, you may wish to place the hanging coil and magnet on an overhead projector.

## N.4. Magnet pendulum swinging over a copper sheet

### Demonstration
A bar magnet tied on a string and swung as a pendulum over a copper sheet induces eddy currents that oppose its motion.

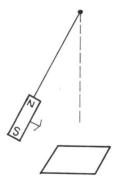

### Equipment
A strong bar magnet; string; and a sheet of copper or aluminum.

### Comment
If you want to show this demonstration on an overhead projector, cut a 3-in.-square piece out of a copper or aluminum sheet and place it on the projector. Suspend the bar-magnet

pendulum from a horizontal support that allows it to swing over the sheet with as little clearance as possible. The amplitude of the swings will decrease rapidly owing to eddy-current braking. The eddy currents will also cause the pendulum to move irregularly as its amplitude decreases. If you remove the metal sheet, the pendulum swings persist without appreciable energy loss. The same is true if you use a metal sheet that has many parallel slots.

## N.5. Dropping a magnet down a pipe

### Demonstration
Induced eddy currents slow the descent of a magnet dropped down a pipe, reducing its acceleration to zero in the case of a very strong magnet. The terminal velocity reached by a strong magnet is proportional to its weight and to the resistivity of the pipe, and it also depends on the pipe's diameter and wall thickness.

### Equipment
A small disk-shaped magnet; a nonmagnetic disk whose mass is equal to that of the magnet; a six-foot-long piece of aluminum pipe, with an inner diameter slightly larger than the diameter of the magnet; and two six-foot-long pieces of copper pipe. One copper pipe should have the same inner diameter and thickness as the aluminum pipe, and the other should have twice the diameter. For best results, the magnet should be as strong as possible for its weight. Half-inch-diameter neodymium magnets with an exceptional strength-to-weight ratio can be obtained from Nakamura, Inc. for about $20.00. Low terminal velocities also require the pipes to have a substantial wall thickness (low electrical resistance). A thick-walled aluminum pipe may give a lower terminal velocity than a thin-walled copper pipe, despite the latter's higher conductivity. The kind of copper pipe used for plumbing applications probably has too small a wall thickness. You may wish to drill ¼-in.-diameter holes through each pipe every 12 inches along their length, in order to observe how soon the magnet reaches its terminal velocity as it moves down the pipe.

### Comment
A magnet dropped down a conducting pipe induces eddy currents whose magnitude depends on the magnet's speed and strength. If the magnet is sufficiently strong, it does not

have to accelerate to a high speed before the induced eddy currents create enough magnetic force to balance the magnet's weight. At this point the magnet descends at a constant terminal velocity, much as it would if acted on by air resistance, another velocity-dependent force.

You can observe the magnet's motion through the pipe as it passes each of the holes along the pipe's length. If you use a magnet with a very high strength-to-weight ratio and if the pipe has thick walls, you should observe that, following an initial acceleration, the magnet travels slowly down the pipe at constant speed, and it passes the holes at equally spaced times. A strong magnet may reach terminal velocity long before it passes the first hole. A weaker magnet probably won't attain a terminal velocity after falling six feet, although you may be able to observe its slowed descent even without a timer if you compare the magnet's time of fall through the pipe with that of a nonmagnetic object. If you are using a strong magnet, you can observe how its terminal velocity depends on its weight, and on the resistivity, diameter, and wall thickness of the pipe:

**1.** Resistivity. Drop the magnet down an aluminum pipe, and down a copper pipe with the same diameter and wall thickness. Find the terminal velocity for each case by timing the magnet's descent through the lower portion of the pipe (past the hole where it seems to stop accelerating). If the terminal velocity is low, you could simply measure the speed for the magnet's descent through the entire pipe. You should find that the terminal velocity for aluminum is 1.6 times that for copper—the ratio of the resistivities of the two metals. This is, of course, not the case if the pipes have different diameters or wall thicknesses.

**2.** Weight. Tape the magnet to a nonmagnetic disk of equal mass, thereby doubling the mass while keeping the strength of the magnet constant. Drop the magnet down the pipe that gave the lowest terminal velocity. You should find that doubling the magnet's mass doubles its terminal velocity.

**3.** Diameter. Drop the magnet down the wide copper pipe. You should find that the magnet has a larger terminal velocity than it had in the narrow copper pipe.

**4.** Wall thickness. Drop the magnet down two pipes that have the same diameter and different wall thicknesses. You should find that the terminal velocity is lower for the thicker-walled pipe.

In order to explain these observations, we begin with $BLv$, the formula for the induced EMF for a wire moving through a magnetic field. (The same expression applies in the case of a stationary wire being approached by a magnet. Thus, the EMF induced in a circular cross section of the pipe being approached by a descending magnet with velocity $v$ and magnetic field strength $B$, is proportional to $B$ and $v$.) The current induced in the pipe depends on its resistance, so it is therefore proportional to $Bvt/\rho$, where $\rho$ is the resistivity of the pipe and $t$ is its thickness. Likewise the magnetic force on the descending magnet, which is proportional to the induced current, is similarly proportional to $Bvt/\rho$. When the magnet reaches its terminal velocity, the magnetic force and weight must balance: $mg = CBvt/\rho$, or $v = mg\rho/CBt$, where $C$ is a proportionality constant. As shown by this last equation, the terminal velocity of the magnet is directly proportional to its mass and to the resistivity of the pipe, and inversely proportional to the pipe's wall thickness.

The dependence of terminal velocity on the diameter of the pipe is more complex, but we should expect wider pipes to have a greater terminal velocity, because less magnetic flux intersects the walls to induce an EMF. The fact that the terminal velocity is predicted to be zero when the resistivity is zero makes it easy to see why magnets can float above superconducting materials.

## N.6. Motor driver and slave

### Demonstration
Turning one DC motor by hand drives a second one connected to it, demonstrating that motors can also function as generators (and vice versa), because of Lenz's law.

### Equipment
Two small DC motors (a "driver" and "slave" combination). They are obtainable from Edmund Scientific Company, for example, for under $10.00.

### Comment
Motors that have permanent magnets also function as generators. If the leads of one motor are connected to another, the "back EMF" generated when one motor is turned drives the second motor. For a motor without a permanent magnet, however, the hand rotation of the armature coil causes no change in magnetic flux, and hence no induced EMF. A similar demonstration can be given by turning one of two hand-powered generators that are connected together.

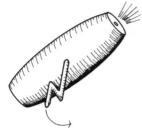

# N.7. Hand-powered generator

**Demonstration**

By cranking a hand-powered generator, you can light a bulb, illustrating induced EMF's and the conversion of mechanical energy into electrical energy.

**Equipment**

A hand-powered generator (available from Nakamura, Inc. for about $20.00); a miniature socket; and a 6 V light bulb. An even less expensive alternative would be a hand-powered flashlight, available from Toys-R-Us for about $5.00.

**Comment**

The flashlight is a small generator; the harder you squeeze the grip, the brighter it lights the bulb. The bulb remains lit only while the grip is pumped. Disconnecting the bulb—which makes it far easier to pump the handle—shows that mechanical work is necessary to produce the electricity that lights the bulb. The bulb is easier to remove, however, when using the hand-cranked generator instead of the hand-powered flashlight. The brightness of the bulb depends, of course, on how rapidly you turn the handle on the generator. You can burn it out if you turn the handle too fast.

# Polarization and Electromagnetic Waves

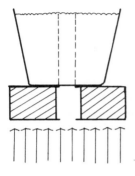

## O.1. Polarization of scattered light

**Demonstration**
The light from an overhead projector is seen to be polarized when scattered by 90° in a slightly milky solution.

**Equipment**
A clear plastic storage box (suggested size 6×6×6 in.); a Polaroid filter; two pieces of opaque cardboard with 1-in.-diameter holes; and some powdered milk or dairy creamer. Polaroid filters can be obtained in sheet form from Jerryco, Inc. or Edmund Scientific Company.

**Construction/Comment**
Fill the clear plastic storage box with water and add a pinch of dairy creamer, so that a beam of light passing through the water becomes visible. If you can't see through the entire box you have added too much powder. Rest the box on a couple of supports on an overhead projector and place opaque cards with 1-in.-diameter holes above and below the supports, so that the holes line up with each other across the gap between the supports. The light passing through the two holes should produce a bright column clearly visible inside the water. For best visibility the column should be located near the front of the box rather than near the middle. Rotate a large Polaroid sheet in front of the light column. If you look directly through the sheet, you will see the light column disappear for one orientation of the Polaroid filter, showing that the scattered light is 100 percent polarized for a 90° scattering. If you are giving the demonstration to a large group, you may need to reposition the Polaroid filter for different parts of the audience to keep it approximately perpendicular to their line of sight.

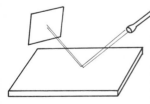

## O.2. Polarization of reflected light

**Demonstration**
Polarization upon reflection is shown by reflecting a light beam off a glass plate and through a Polaroid filter.

**Equipment**

A Polaroid sheet; a high-intensity focusable flashlight or slide projector; and a glass plate. The glass surface of an overhead projector works fine.

**Comment**

Reflect a light beam off the glass plate onto a screen. When the Polaroid sheet is interposed in the beam and rotated, the degree of polarization of the beam is shown by the extent of the intensity variation of the image on the screen. For a completely polarized beam, the image on the screen fades out entirely for some orientation of the Polaroid sheet. Complete polarization should occur when the angle of incidence of the light equals Brewster's angle, $\tan^{-1}n$, which for glass is about 56°, making the angle between the light beam and the glass surface 34°.

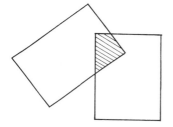

## O.3. Crossed Polaroids and optical activity

**Demonstration**

By rotating one of two overlapping Polaroid filters on an overhead projector, and observing the light intensity, you can observe a variety of polarization effects.

**Equipment**

Two Polaroid sheets; a U-shaped piece of ⅛-in.-thick Lucite; a jar of Karo syrup sugar solution; and a plastic cup.

**Comment**

First, see how the light intensity drops to virtually zero for one relative orientation of the two Polaroid sheets ("crossed Polaroids"). Next, see the effect of inserting a third sheet between two crossed Polaroids. Surprisingly, some light passes through the three sheets even though it didn't pass through the two sheets. This is easily explained by considering the three vectors in Plate O.3a, which represent the polarization directions of sheets 1, 2, and 3. Vectors 1 and 3 (for the crossed Polaroids) are perpendicular, because no light passes through two sheets with these polarization directions. But now suppose sheet 2 is inserted between sheets 1 and 3. Because the polarizations of 1 and 2 are not perpendicular, a certain fraction of the light that passes through sheet 1 also passes through sheet 2. This light has a polariza-

**O.3a** Polarization vector directions for three Polaroid sheets

tion direction represented by vector 2. Likewise, a nonzero fraction of this light passes through sheet 3, since vectors 2 and 3 are not perpendicular to each other. If you change the order of sheets 2 and 3, of course, no light passes through. The insertion of sheet 2 between 1 and 3 acts to rotate the direction of polarization of light emerging from sheet 1 from the direction of vector 1 to that of vector 2.

Several other easily demonstrated methods exist for rotating the plane of polarization. For example, place a U-shaped piece of Lucite between two Polaroid sheets whose polarization directions are at right angles to each other. If you squeeze the top ends of the U together, those regions of Lucite under internal stress cause a rotation of the polarization direction, and create a pattern that depends on the applied force and on the exact shape of the Lucite piece. This effect allows engineers to use plastic models to study stress patterns in objects subject to forces.

Optical activity, another method of rotating the plane of polarization, occurs in sugar solutions, such as Karo syrup. The amount of polarization rotation in an optically active medium depends on the thickness of the medium and on the wavelength of the light. Thus if you add Karo syrup to a plastic cup placed on an overhead projector, with one Polaroid sheet below the cup and one above, you see a color that depends both on the height of Karo syrup in the cup and on the angle between the two Polaroid sheets. Different colors occur because the polarization direction is rotated by different amounts for different wavelengths. Therefore, the amount of transmission of a given wavelength through the second Polaroid filter depends on both the height of the syrup and the filter's angle.

## O.4. Wave on a spring through a vertical slit

**Demonstration**
Shaking a long spring that passes through a vertical slit demonstrates the concept of polarization of transverse waves.

**Equipment**
A 6-ft-long spring, and a vertical slit barely wide enough to allow the spring to pass through.

**Comment**
Hold one end of the spring and pass the spring through the slit, attaching the other end of the spring to some fixed object. It is important that the slit be made sturdy enough that

it will not be knocked over when you start shaking the spring from side to side. One makeshift arrangement that seems to work fine is two stacks of heavy physics textbooks arranged at 45° angles, so that the spring can be moved up and down in the space between their corners. You can then show that waves travel on the spring through the slit when it is shaken up and down, but not when it is shaken side to side. (On the other hand, you can get side-to-side *standing* waves through the slit if the frequency of shaking and the distances work out to make the position of the slit a node.) In order to illustrate how polarization works you would like to be able to send an "unpolarized" wave into the slit and see what comes out. Instead of an unpolarized wave, what you can do is send in a circularly polarized wave (by moving your hand in a circle), and show that the wave that passes through the slit possesses only the vertical polarization component.

## O.5. Calcite crystal

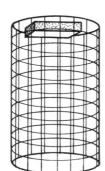

**Demonstration**
A piece of doubly refracting calcite crystal shows a double image of whatever is placed beneath it; the two images correspond to crossed polarizations.

**Equipment**
Calcite crystal (obtainable from geological supply companies, such as Ward's Natural Science Establishment, Inc.); a transparency placed beneath the crystal on an overhead projector; and a Polaroid filter.

**Comment**
When you rotate the calcite crystal on top of a fixed transparency, one image—formed by the so-called "ordinary rays"—stays fixed, and the other image—formed by the "extraordinary rays"—rotates. The ordinary and extraordinary rays are polarized along mutually perpendicular axes. The fact that these two images have different polarizations can be shown by placing a Polaroid filter on top of the calcite crystal and rotating it.

## O.6. Shielding a radio

**Demonstration**
Wrapping a portable radio in transparent wire mesh to eliminate the reception is just as effective as wrapping the radio

in opaque aluminum foil, showing that radio waves must have a longer wavelength than light waves.

**Equipment**
A small portable radio; some aluminum foil; some fine-mesh wire screen; and some coarse-mesh chicken wire. Shape the chicken wire into a cylindrical cage with a bottom but no top, into which the radio can be placed.

**Comment**
Light waves clearly penetrate the open spaces of the fine-mesh wire screen because we can see through it, but light waves cannot penetrate the spaces between the atoms in the aluminum foil, which is opaque. Radio waves, however, are blocked just as effectively by both aluminum foil and fine-mesh wire screen, as you can easily confirm by wrapping a portable radio in each.

If the radio is now placed in a coarse-mesh chicken-wire cage, the signal is only weakened, and not completely cut off. The extent of signal attenuation depends strongly on the radio's orientation. This should not be interpreted to mean that radio wavelengths are about the size of the chicken wire spaces, since no simple relationship exists between the wavelength and the size of the mesh waves can pass through. Some signal gets through the chicken wire, even though radio wavelengths are much longer than the open spaces, because the metal wires act like a receiving antenna that absorbs and re-radiates the radio signal with a 180-degree phase reversal, and this signal interferes with the original incident wave. The destructive interference is complete in the case of the aluminum foil or the fine-mesh wire screen, but only partial for the coarse-mesh chicken wire.

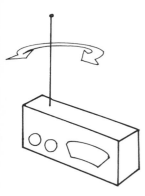

## O.7. Transverse nature of radio waves

**Demonstration**
By pointing the antenna of a portable AM radio in different directions, you can show that the radio waves are transverse.

**Equipment**
A portable radio that has a linear extendable antenna.

**Comment**
Radios exhibit some puzzling directional properties that can be explained by the transverse nature of electromagnetic waves, and by an understanding of the two types of antennas

found in portable radios. Tune the radio to a particular AM station, and observe the change in reception with the antenna extended and withdrawn. You will probably observe very little change, because except for car radios, most portable radios receive their signal primarily through an internal loop antenna (which responds to the magnetic field of the radio wave), rather than through the linear antenna (which responds to the electric field).

Place the radio on a table with the linear antenna vertical, and slowly rotate the radio about a vertical axis. You should find that the signal goes to zero for two orientations that are 180° apart. At these orientations, the plane of the internal loop antenna is perpendicular to the direction of the transmitting station. The absence of a signal at these orientations is explained by the fact that no magnetic flux from the transverse electromagnetic wave passes through the plane of the loop. The demonstration will probably fail to work if you repeat it using an FM station, because the circuitry for FM reception includes an automatic gain control that boosts the signal internally to compensate for variations in signal strength.

# Section P

## Geometrical Optics

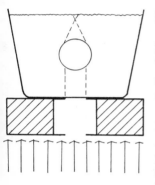

### P.1. Underwater optics

**Demonstration**

The optical properties of lenses, prisms, and mirrors can be conveniently demonstrated underwater using a slightly milky solution and the light from an overhead projector.

**Equipment**

A clear plastic storage box (suggested size 6×6×6 in.); some powdered milk or dairy creamer; a variety of strong lenses, mirrors, prisms, Lucite or glass rods, test tubes or plastic vials; and a small rectangular plastic box.

**Construction**

Fill the clear plastic storage box with water, and add a pinch of dairy creamer so that a light beam passing through the water becomes visible. If you can't see through the entire box, you have added too much powder. Rest the box on a couple of supports on an overhead projector and place opaque cards with 1-in.-diameter holes above and below the supports, so that the holes line up with each other across the gap between the supports. The light passing through the two holes should produce a bright column clearly visible inside the water. For best visibility the column should be located near the front of the box rather than near the middle.

**Comment**

Try placing each of the following underwater in the vertical column of light:

    **(a)** strong glass lenses or Fresnel lenses, singly and in combination;

    **(b)** glass or Lucite cylinders of various radii;

    **(c)** air-filled plastic cylinders, or glass test tubes;

    **(d)** air-filled plastic rectangular boxes;

    **(e)** convex, concave, and plane mirrors.

The focal length of a lens is inversely proportional to the difference in indices of refraction between the lens and the surrounding medium. Hence, glass or Lucite lenses have focal lengths underwater that are much longer than the focal

177

lengths they have in air—about three times longer for lenses that have an index of refraction of 1.5. Therefore, you need to use lenses or cylinders with very short focal lengths, in order that the focal point remain underwater. Test tubes make good air-filled convex lenses, which are readily seen to be *diverging* when placed under water. The submerged air-filled plastic box acts like a prism, and it exhibits total internal reflection for certain orientations. The convex and concave mirrors behave underwater just as they do in air.

## P.2. Multiple images between two inclined mirrors

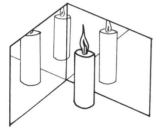

### Demonstration
A candle or a light bulb placed between two mirrors inclined at an angle gives rise to multiple images whose number depends on the angle.

### Equipment
Two square mirrors (12×12-in. mirror tiles are particularly good); a candle or other light source; duct tape; and polar-coordinate graph paper.

### Comment
When the mirrors are taped loosely together, their angle can easily be varied and they will stand up without additional support. If they are viewed from a level height, the number of multiple images of an object placed in the space between the mirrors increases as the angle between the mirrors decreases. The number of images increases by one, each time the mirror angle is decreased below one of the values 360°/ $N$, where $N = 2, 3, 4, \ldots$ Thus, the number of images as a function of angle between the mirrors is as follows:

| Angle in degrees | Number of images |
|---|---|
| 120 to 180 | 3 |
| 90 to 120 | 4 |
| 72 to 90 | 5 |
| 60 to 72 | 6 |
| 51.4 to 60 | 7 |

In general, we can write the number of images $N$ as an explicit function of the angle $A$: $N = \text{Int}(360/A + 1)$, where the "Int" function rounds the quantity in parentheses to the nearest lower integer. According to this equation, the number of images becomes infinite as the angle *A* between the mirrors approaches zero. When you stand between large parallel mirrors you can see this effect. In practice, however, only a finite number of images can be seen because they get progressively fainter with each mirror reflection.

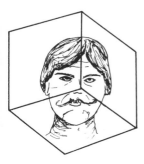

## P.3. Three-corner reflector

### Demonstration
When looking at a three-corner reflector (the interior corner of a cube), you see yourself no matter what angle you view the reflector from.

### Equipment
Three square mirrors whose edges are taped together using duct tape so that they are all mutually perpendicular. (12×12-in. mirror tiles sold in home-improvement stores would be one possibility.)

### Comment
Light rays incident on the mirrors from any direction bounce off one, two, or three mirrors, and emerge in the direction opposite to the direction they came from, owing to the 90° angle between the mirrors. As a consequence you will be able to see your image from any angle at which you can see the mirrors. This is the basis of the red reflectors and special reflecting strips, which have many tiny three-corner reflectors on their surface.

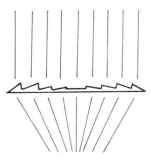

## P.4. Fresnel lens

### Demonstration
A large plastic Fresnel lens can be shown to have the focusing property of a conventional lens.

### Equipment
A Fresnel lens (available from Jerryco, Inc. or Edmund Scientific Company).

### Comment
Place the large Fresnel lens in front of your face and others will see you magnified. Put it on an overhead projector and the transparent lens looks completely opaque except for a bright spot in its center, which is the out-of-focus image of the projector bulb. Adjust the projector to focus the bulb's

image. You can determine the lens' focal length by focusing the light from a distant window onto a sheet of paper and measuring the distance from the lens to the paper. If you have a second Fresnel lens, you can then show that the two lenses combined have half the focal length (twice the power) of one alone.

## P.5. "TV rock" (ulexite)

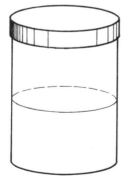

### Demonstration
Ulexite, a naturally occurring mineral that consists of many parallel optical fibers, transmits images across the width of the material.

### Equipment
Ulexite (which can be obtained from geological supply companies such as Ward's Natural Science Establishment, Inc.).

### Comment
Place the ulexite or "TV rock" on top of a written page, and the lettering appears on the top surface of the rock. The image of the lettering is transmitted through many closely spaced optical fibers. You can easily show this on an overhead projector by placing the TV rock on top of a transparency that has lettering on it, and tilting both the transparency and the rock a bit to show that the lettering appears on the top surface.

## P.6. Total internal reflection at a two-liquid interface

### Demonstration
Total internal reflection occurs when the interface of two immiscible liquids is viewed from above, if the liquid on top has the higher index of refraction.

### Equipment
A pair of immiscible liquids (corn oil and a water-alcohol mixture) in a jar or jug; and some food coloring.

### Construction/Comment
Add some food coloring to a water-alcohol mixture, and pour corn oil on top of the mixture. If less than a certain

alcohol-to-water ratio is present (about 2 : 1 by volume), the oil will be the lighter liquid and form the upper layer. The interface between the two layers will appear to be shiny (totally internally reflecting) because the oil has the higher index of refraction. Thus the color of the lower liquid will not be seen when looking at the interface. If we repeat the demonstration with the oil added to a water-alcohol mixture that has a somewhat higher alcohol-water ratio, the oil will sink to the bottom, and the interface will no longer appear to be totally reflecting. In this second case do not add food coloring to the water, since the upper layer should be transparent. Depending on temperature and humidity, the oil layer may become temporarily cloudy because fat precipitates out of solution. This can also occur if the mixture is vigorously shaken, in which case the cloudiness will persist.

The two immiscible liquids in a jar can also be used to show extremely slow standing waves, especially if the alcohol-water mixture has almost the same density as the oil. With the jar filled to the brim and sealed, create a standing wave by momentarily tilting the jar. The resulting standing wave has an extremely slow velocity because the wave velocity at a two-liquid boundary layer is proportional to the difference in the liquid densities.

## P.7. Fountain of light

**Demonstration**
Due to total internal reflection, light can be trapped in a water stream flowing out of a hole in a cup, but it emerges at a point where the stream begins to break up.

**Equipment**
A high-intensity focusable flashlight; and a plastic cup with an 1/8-in.-diameter hole in its side near the bottom.

**Comment**
Fill the cup with water, allowing the water to flow out the hole, and shine the flashlight toward the hole from a point directly across, on the other side of the cup. (The curved walls of the cup are a benefit in this demonstration rather than a detriment, because they act to focus the flashlight beam, causing a greater light intensity at the point of the hole than a container with flat sides would.) If the room is darkened, the light is seen to remain within the water stream up to the point where the stream breaks up. Up to that point, the

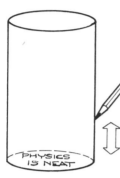

angle of incidence of the light exceeds the critical angle, so the light is trapped in the smoothly curving water stream. Observe how the change in shape of the water stream, as the water level descends in the cup, affects the point along the stream at which the light emerges.

## P.8. Apparent depth of a submerged object

### Demonstration
The apparent depth of the bottom of a water-filled jar can be found by focusing an overhead projector on lettering beneath the jar, and sliding a pencil along the side of the jar until the pencil point is in sharp focus.

### Equipment
A pencil; and a tall jar with a flat clear bottom (such as a 12-in.-tall spaghetti holder).

### Comment
Fill the jar to the brim with water and place it on top of a transparency (on which words are written) on an overhead projector. Focus the projector on the transparency as carefully as possible, and then move the pencil along the side of the jar so that the pencil tip is also in sharp focus. The depth of the pencil below the water surface is the measured apparent depth of the bottom of the jar. You should find that the apparent depth is approximately equal to the actual depth divided by 1.33, the index of refraction of water. Alternatively, you could use this demonstration to find the index of refraction, $n$, from the equation $n = d/a$, the ratio of the actual and apparent depths. You can also obtain an uncertainty in the measurement by observing the limits on the apparent depth, i.e., the highest ($a_1$) and lowest ($a_2$) pencil locations for which the pencil tip remains in sharp focus. The measured index of refraction can then be expressed as $n \pm dn$, where $n = 2d/(a_1 + a_2)$ and $dn = \frac{1}{2}(d/a_2 - d/a_1)$, which may be compared with 1.33, the index of refraction of water.

The hardest part in doing the demonstration is focusing the projector to show the lettering under the jar as clearly as possible. Turning the knob on the projector slightly disturbs the water surface, and makes it difficult to achieve the best focus. This difficulty is removed if a solid transparent medium such as glass is used. In addition, using a medium such as glass, whose index of refraction is higher than that of water, produces a more pronounced effect. But the thickness of the medium is also important, and it is difficult to find a

**Geometrical Optics**

piece of glass six or more inches thick. (One possibility would be to use a long glass or plastic rod with polished ends, or else a glass plate with polished edges.)

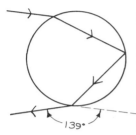

## P.9. Rainbow formation

### Demonstration
Rainbows caused by internal reflections in spherical water drops can be produced using a water-filled jar illuminated by a flashlight.

### Equipment
A cylindrical glass jar; and a focusable high-intensity flashlight. To obtain best results, the thickness of the glass should be as uniform as possible to minimize optical distortion. A beautiful rainbow suitable for viewing by a large group can also be shown using a water-filled plastic cup on an overhead projector. Still another demonstration of rainbow formation uses "rainbow dust" (the tiny glass spheres that are used by highway departments to put a coating on road surfaces that retroreflects headlight beams). See the discussion in Frank S. Crawford's article in the *American Journal of Physics 56(11)*, 1006 (November 1988).

### Comment
Primary rainbows when light enters spherical water drops, and emerges after one internal reflection. Even though rays incident at different points on a spherical drop emerge at different angles, it can be shown that for yellow light a bunching occurs when the emerging rays make an angle of about 139° with respect to the incident direction. Colors are seen because the variation of index of refraction with wavelength causes light of different wavelengths to emerge at slightly different angles.

You can observe the 139° reflection by shining a focused flashlight through the side of a water-filled jar in a darkened room. For best results, place the jar on a piece of white paper, and aim the flashlight beam slightly downward, to leave light streaks on the paper. Starting with the beam next to the middle of the jar, slowly move the flashlight toward the edge of the jar without changing the direction of the beam. This motion causes the light to make larger and larger angles of incidence with the jar. The light emerging from the jar increases its angle with the incident beam as the flashlight moves toward the edge of the jar. At some particular point, the light beam inside the water-filled jar exceeds the critical angle. The light beam then undergoes a total internal reflection, and it emerges at an angle of 139° with respect to the incident direction.

Although the individual colors in the spectrum are only separated by about 2° from red to violet, you should be able to see that some dispersion has occurred, if the glass distortion is not too large. Unlike real rainbows, which form an arc, the light here occurs along one specific direction, because we are using a cylindrical, not a spherical, volume of water. If you do the demonstration using a spherical volume of water (or glass), it is difficult to obtain a visible rainbow arc, because the light is not concentrated along one specific direction. A visible rainbow can only be obtained with a single sphere if you use a very intense, well-collimated light source. A beautiful rainbow can be obtained using a large number of tiny spheres, however, as described in Frank Crawford's article on "rainbow dust."

One other simple method of obtaining a pretty rainbow arc without a sphere is to place a plastic cup of water on an overhead projector. Refraction of the light from the projector will create a circular rainbow on the ceiling. Upward light rays incident on the outside of the steeply sloped sides of the cup have a large angle of incidence, and therefore when they refract into the water the degree of dispersion is considerable.

## P.10. Turning the world inside out

### Demonstration
The focal length of a convex mirror can be found using a meterstick, and the mirror equation can be checked on an overhead projector.

### Equipment
A reflecting sphere (such as a stainless-steel ball or a Christmas tree ornament); a convex mirror (obtainable from automobile supply stores); and a meterstick.

### Comment
When you view a reflecting sphere, you can see the images of objects in a field of view of nearly 360°. In a sense the reflecting sphere "turns the world inside out," by imaging a point inside the sphere for every outside point. $p$, the distance between the sphere and an object, and $q$, the distance between the sphere and the object's image behind the surface of the sphere, are related through the mirror equation:

$1/p + 1/q = 1/f$, where $f$ is the focal length of the mirror. For a convex mirror with a radius of curvature $R$, the focal length is $-R/2$.

A simple way to verify the two preceding relationships is to hold one end of a meterstick against a reflecting sphere that has as large a radius as you can find. (The Christmas tree ornament would be preferable to a small metal ball.) Center the sphere at the end of the meterstick, which should be oriented radially outward. Hold a pencil against the meterstick, and slide it along the stick's length until it is located at the point at which the image of the pencil appears to be midway between the images of the two ends of the meterstick. The far end of the meterstick is far enough from the sphere that its image distance is nearly equal to the focal length. The end of the meterstick that is in contact with the sphere has an object and image distance of zero. The pencil, which appears halfway between the ends of the meterstick, must have an image distance equal to half the focal length and, according to the mirror equation, an object distance that is also half the focal length. Thus by measuring the pencil's distance to the sphere you can determine $f$, and you can see how well the relation $f = -R/2$ is verified.

A similar demonstration can be done on an overhead projector, but in this case a convex mirror—that has a radius of curvature much larger than that of the sphere—is needed to get a readily observable result. Place the mirror (reflecting side up) on the projector, and focus on the outline of the mirror, which of course is opaque. Change the projector focus so that the bright spot at the center of the mirror is in focus. This spot is the image of the reflection from the top (movable) lens of the projector. The distance of that top lens above the mirror is the object distance $p$, and the distance you had to lower the lens to focus the spot is the image distance $q$ (because if you don't move the lens at all the image must be right at the mirror, with $q = 0$). You can use the measured values of $p$ and $q$, together with the value of $f$ found from the meterstick method, to see how well the mirror equation is verified.

# Interference and Diffraction

*General*

## Q.1. Superposition of circle transparencies

**Demonstration**
Superimposing multiple transparencies of circle patterns
simulates the interference and diffraction of waves from two
or more sources.

**Equipment**
Transparencies made by photocopying the circle patterns of
Plates Q.1a, Q.1b, and Q.1c; and a cardboard transparency
frame.

**Comment**
A superposition of two circle transparencies on an overhead
projector shows the features of the two-source interference
pattern in a quantitative way. Each circle transparency represents the waves from a point source, the spacing between
adjacent circles being one wavelength. Using these circle
transparencies, the following four demonstrations can be
given:

# Interference &
# Diffraction: General

**Q.1a** Transparency template for waves from a point source

Section Q

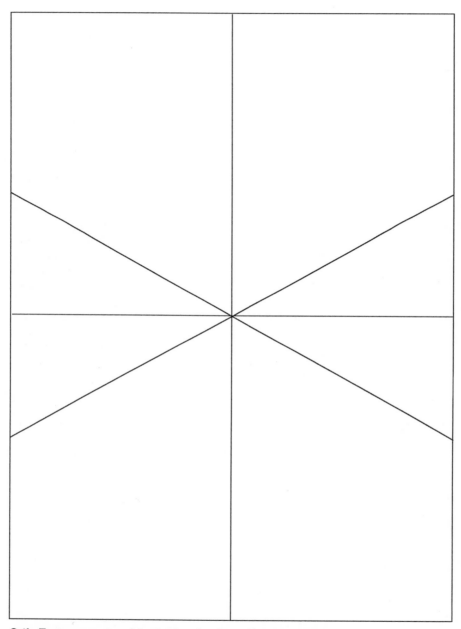

**Q.1b** Transparency template showing directions of interference maxima for $d = 2\lambda$

Interference &
Diffraction: General

**Q.1c** Pattern produced
from the overlay of six
circle transparencies
whose centers are
spaced two wave-
lengths apart along
the vertical axis

(1) For a fixed wavelength $\lambda$, the angular spacing between maxima and minima decreases as the distance between sources, $d$, increases.

(2) The specific angles along which maxima and minima
occur are given by $\sin A = m\lambda/d$ (maxima) and $\sin A =
(m + \frac{1}{2})\lambda/d$ (minima). To demonstrate these relations quan-
titatively, prepare a transparency from a photocopy of Plate
Q.1b, which has lines drawn along directions of maxima and
minima expected for sources separated by two wavelengths:
$d = 2\lambda$. You can directly verify the conditions for maxima
and minima for the $d = 2\lambda$ case by putting two circle transparencies made from photocopies of Plate Q.1a on top of the
one made from Plate Q.1b, and positioning them until the
maxima in the interference pattern coincide with the lines
drawn in Plate Q.1b. You should find that the sources have
a separation of two wavelengths at this point, i.e., the centers
of the two circle patterns are displaced by twice the circle
spacing.

(3) In order to show the interference pattern for two sources that have slightly different wavelengths, you should overlay two transparencies made from Plate Q.1a, one of which is made from a photocopy that is slightly enlarged or reduced relative to the other. In this pattern the maxima and minima lie along curved paths rather than along straight lines. As a result, an observer stationed at a distant point at any given angle would see a continually changing interference phase, since the circle patterns only represent a "snapshot" of the wave pattern at a given instant of time. Thus, observable interference between two sources requires that they have exactly the same wavelength; otherwise, they will produce a time-varying phase at a given point—i.e., beats.

(4) For $N = 3$, 4, 5, and 6 circle transparencies copied from Plate Q.1a, diffraction patterns are produced that show principal maxima at the same angles as was the case for two sources, and $N - 1$ lower-intensity secondary maxima between each pair of principal maxima. Thus the three-source pattern has one faint secondary maximum between each pair of principal maxima, and the four-source pattern has two. In addition, the widths of the principal maxima narrow as more circle transparencies are combined. These features can be seen in Plate Q.1c, a photograph made of a superposition of six circle transparencies. Although it is quite easy to obtain a good two-source pattern, obtaining a good 3-, 4-, 5-, or 6-source pattern requires a great deal of care. The following precautions should be helpful:

(1) Before making transparencies from photocopies of Plate Q.1a, make a mark at the bottom of each photocopy. This will allow you to orient the transparencies so that they all face the same way. Owing to small nonuniformities in the circle patterns, placing one or more upside down will result in irregular patterns, as you can easily verify.

(2) Make each circle transparency from a fresh photocopy of Plate Q.1a; otherwise, they may get progressively fainter.

(3) Make a transparency from Plate Q.1b showing the angles along which maxima are predicted for the case $d = 2\lambda$, and tape it securely to the *underside* of a cardboard transparency frame, to help line up the circle transparencies.

(4) Tape the first two circle transparencies on the *top side* of the cardboard frame, after you have lined up the positions of the maxima so that they accurately coincide with the lines at $0°$, $\pm30°$, $\pm90°$, $\pm150°$, and $180°$ in Plate Q.1b. Align them as accurately as possible, and tape them down securely. One

way to verify the accurate lineup is to see a symmetrical pattern in all four quadrants when the transparencies are pressed completely flat. The two "sources" should be located at positions one wavelength above and below the center of the pattern, in order that the maxima appear at their proper angles.

(5) Put the third circle transparency on top of the first two and move it around until (a) equally bright principal maxima appear at the same angles as before in *all four quadrants*, and (b) equally bright secondary maxima appear midway between the principal maxima in *all four quadrants*. Don't tape the third transparency down until you have it aligned as accurately as possible and have made it completely flat. The center of the third circle transparency should be located two wavelengths from one of the other two centers, and all three centers should lie on the *y* axis when the proper pattern is achieved. Very small movements will result in appreciable changes in the pattern.

(6) Repeat step (5) for 4, 5, and 6 circle transparencies. Do not tape each transparency down until it has been aligned as accurately as possible using the four-quadrant symmetry test. It may be difficult to see the secondary maxima after four transparencies have been superimposed, as they get progressively fainter. The transparencies must all line up the same way if you are to avoid the irregularities noted earlier.

Notice the extreme narrowness of all but two of the principal maxima in the pattern for six superimposed circle transparencies (Plate Q.1c). The broadness of the maxima along ±90° is a consequence of the fact that the directions between successive minima involve steps of equal size in sin *A*, not in *A*. Some faint secondary maxima can also be seen between principal maxima in Plate Q.1c.

## Q.2. Moire patterns

### Demonstrations
Moire patterns—essentially an interference effect—can be shown for a variety of cases.

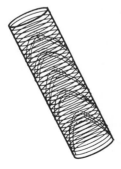

### Equipment
A piece of silk fabric folded over on itself; and a coiled spring. An inexpensive collection of moire patterns ("Moire Array") is available from Edmund Scientific Company for about $3.00.

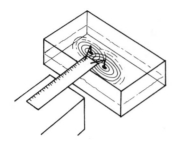

**Comment**
Moire patterns can be generated using two pieces of silk, or using a spring that is stretched and/or rotated. The interference and diffraction patterns obtained with the circle transparencies used in Q.1 are also examples of moire patterns.

## Q.3. Ripple tank

**Demonstration**
A simple ripple tank and variable-frequency wave source can be constructed to illustrate interference and diffraction of water waves.

**Equipment**
A water tank; a clear plastic ruler (stiff type); a block of wood; a comb; some paper clips; and clay. A suitable water tank would be either a clear plastic storage box or a one-piece plastic box picture frame; the latter is preferable even though it can be filled only to depths of an inch or less. The water tank should be as large as possible, while still being stable on an overhead projector. A 12×15-in. plastic box picture frame with a 1½-in. depth is quite suitable. The wooden block should be of a height such that a ruler resting on it just projects over the edge of the water tank.

**Construction/Comment**
Make a variable-frequency source of water waves by straightening a paper clip, taping it to the end of the plastic ruler, and putting a small clay ball on the end of the paper clip. (Bend the end of the straightened paper clip into a right angle or the clay ball will fall off.) Put the storage box on an overhead projector and add cool water. (If the water is too warm the clay ball will soften and fall off.) The water should be deep enough that when the ruler rests on the wooden block and projects over the edge of the box (without touching the box), the clay ball's center lies right on the water surface.

The box picture frame makes a better water tank than the deeper plastic box does because the box picture frame's shallower depth allows the clay balls to reach the water surface more easily when the water depth is only ¼ inch. The advantage of using such a shallow water depth is that the wave velocity is then reduced, and the wave pattern becomes easier to see, particularly when you are viewing interference minima and maxima.

If you pluck the end of the ruler, circular waves will persist as long as the ruler oscillates. To vary the frequency of the oscillations, you simply vary the length of ruler projecting over the end of the block. To produce two-source interference patterns, make a double wave source, using two paper clips and clay balls on the two edges of the ruler, with a source separation of about an inch. It is important to make the two clay balls the same size and the same distance below the ruler, so that both balls enter the water at the same time and to the same extent.

Producing good wave interference patterns is largely a matter of technique, which includes all of the following:

(1) Be sure that the water has the correct depth—both of the clay balls should rest on the surface half submerged.

(2) The fixed end of the ruler must be held firmly against the wooden block that is resting on the overhead projector. If the end of the ruler is not planted firmly, the oscillations of the free end will not persist.

(3) Pluck the free end of the ruler with one finger, causing *small* amplitude oscillations that do not cause the clay balls to either leave the water or become completely submerged. Think of the water surface as a membrane (which it is!) that you want to disturb but not penetrate. Pluck the ruler repeatedly, slightly varying the amplitude each time until you get the best-looking waves. Let the water surface settle between plucks.

(4) The two clay balls should be no closer than 1–2 inches away from any wall of the box, to minimize wave reflections from the walls. It would be best to have the balls placed at the center of the storage box, so that they are equally distant from all walls.

(5) Vary the frequency of the oscillations by varying the length of ruler hanging beyond the edge of the wooden block. Varying the frequency causes both the wavelength and the wave velocity to vary. Changing the wavelength in turn affects the angles along which maxima and minima in the two-source interference pattern occur, and affects the visibility of the pattern as well.

(6) Finally, and most importantly, focus the overhead projector on the ruler, and not on the water surface. Strangely, this makes the waves easier to see. You see the waves in the first place because the changing curvature of the water surface acts like a lens of different curvature at each point. Having the water surface out of focus compensates for the water lens at a given point on the wave. An additional benefit of focusing on the ruler is that you

can read the ruler and estimate the wavelength more easily.

Once you have good two-source interference patterns, observe at which angles maxima and minima appear. The minima, which are generally easier to see, should occur at $\sin A = \pm(m + \frac{1}{2})\lambda/d$, and the maxima at $\sin A = \pm m\lambda/d$, where $m$ is an integer, $\lambda$ is the wavelength, and $d$ is the source spacing. For example, according to these equations, the predicted angles for maxima when $\lambda = d$ are $A = 0°$, $\pm 90°$, and $180°$, while if the wavelength is halved ($\lambda = d/2$) the predicted maxima are at $A = 0°$, $\pm 30°$, $\pm 90°$, $\pm 150°$, and $180°$. As this example shows, the angular spacing between maxima (and minima) decreases as the ratio $\lambda/d$ decreases.

A simple way to observe the decrease in angular spacing with decreasing wavelengths is to produce interference patterns with progressively smaller amounts of ruler overhanging the wooden block. With somewhat greater difficulty, you can also check that the maxima and minima occur at the predicted angles for waves of a given wavelength. Place the transparency made from Plate Q.1c of demonstration Q.1 (which has lines at $0°$, $\pm 30°$, $\pm 90°$, $\pm 150°$, and $180°$) underneath the water-filled box. The transparency should be properly oriented with respect to the two wave sources (the sources should lie on the $90°$ line and straddle the origin). Vary the amount of ruler overhang (thereby varying $\lambda$ for fixed $d$) until you produce interference maxima along directions that match those in the transparency. Estimate the wavelength of the rapidly disappearing waves as well as you can, and see how well the condition $\lambda = d/2$ is satisfied.

To demonstrate the diffraction of waves from a row of $N$ sources, make an $N$-source generator out of a five-inch pocket comb. Remove teeth from the comb, leaving six teeth (including the two ends) that are approximately one inch apart and are equally spaced. Mount small clay balls of equal size on these six remaining teeth, and tape the comb to the end of the ruler, making sure that the clay balls all rest half submerged on the water surface.

For the six-source pattern, the directions of the maxima are the same as for two sources, but have better definition. In other words, much lower wave intensity should be observed in directions away from the maxima. Thus for six sources you should observe nearly plane waves traveling along the directions determined by the condition for principal maxima for the $d = 2\lambda$ case, although you may not see anything along the $\pm 90°$ directions because the principal maxima are

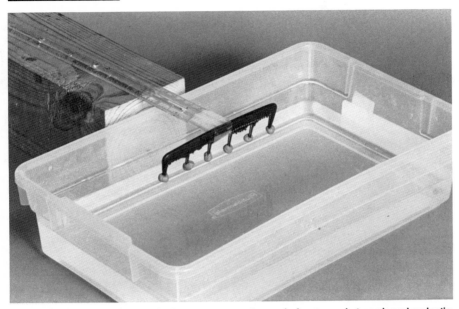

**Q.3a** Ripple-tank apparatus with six-source generator made from a pocket comb and a plastic ruler

much broader along these directions (as shown in demonstration Q.1). To see the ±30° plane waves best, rotate the comb so that it makes a 30° angle with respect to the walls of the box. In this case, the first nonzero-angle principal maximum would be expected to travel parallel to the box walls. (It is important to have the comb in the center of the box for the six-source case.)

If the number of sources, $N$, is sufficiently large (try a comb with twelve teeth spaced ½ inch apart), the $N$ waves when combined approximate a plane wave. You can observe the diffraction of such plane waves through an aperture by placing a barrier with openings of various sizes in front of the source. You should find that the extent of diffraction increases as the aperture size decreases, or as the wavelength increases.

*Acoustic*

## Q.4. Sound through a fan

**Demonstration**

The dependence of diffraction on wavelength explains why only high frequencies are modulated when sound passes through a fan.

Section Q

**Equipment**
A small fan, and a Radio Shack beeper (see demonstration J.6).

**Comment**
If a rotating fan is placed on the straight line between a source of sound and a listener, the fan blades periodically cut across that straight line. Sound, like any wave, diffracts around obstacles to an extent that depends on the ratio of the wavelength to the size of the obstacle. For a relatively low-frequency sound, such as the sound of your voice, the diffraction is appreciable, and little effect should be heard by the listener when you talk through the fan, especially if you have a deep voice. On the other hand, if you place the high-frequency beeper (3600 Hz, $\lambda = 4$ in.) behind the fan, much less diffraction occurs, and the sound should be more nearly cut off during those times when the fan blades cut across the straight line between the beeper and the listener. The listener will therefore hear the beeper sound modulated by the fan's rotation. This produces a warbling kind of sound, one that is quite distinct from the sound of the beeper without the fan.

## Q.5. Diffraction with a rolled-up carpet

END CAP

**Demonstration**
By producing sounds through rolled-up carpet cylinders of several diameters, you can show how the extent of diffraction depends on the ratio of wavelength to cylinder diameter.

**Equipment**
Pieces of carpeting to make two 1-ft-long rolled-up cylinders, with diameters of 1 foot and 4 inches, respectively; a piece of aluminum sheet (obtainable at home-improvement or hardware stores); duct tape; and a Radio Shack beeper (see demonstration J.6). You can obtain carpet samples at a carpet store.

**Construction**
Make each rolled-up cylinder with the soft (sound-absorbing) side of the carpet on the inside. Use a long enough piece of carpet to ensure that the 1-ft-diameter cylinder has sufficient structural strength. Use duct tape to keep the carpeting rolled up.

You will also need to make an end cap, which will keep the sound from leaving the back of the cylinder and also serve as a mounting for the beeper; the cap should be circular or square and completely cover one end of the cylinder. To keep the end cap on the cylinder, cut a 2-in.-wide strip of aluminum sheet, and shape it into a hoop whose radius equals the inner radius of the cylinder. Securely tape the hoop to the end cap. When you place the end cap on the cylinder, this hoop fits snugly inside the cylinder and keeps the end cap in place even when the cylinder is held in a horizontal position.

Cut another 2-in.-wide strip of aluminum, and with it make a circular enclosure at the center of the end cap into which the beeper can be placed snugly. Tape this enclosure to the end cap. Repeat this entire process to make the 4-in.-diameter rolled-up carpet cylinder and end cap. The beeper should be attached to a 9 V battery and wire as explained in demonstration J.5. (Put tape over the battery terminals so you don't short it out when placing it in the aluminum enclosure.)

**Comment**

Place the beeper snugly in the end-cap enclosure of the 12-in.-diameter carpet cylinder. Put the end cap on the cylinder and activate the beeper by connecting the wires attached to it and to the battery. Sweep the direction of the carpet cylinder horizontally, so that it points toward the listener(s) at the middle of the sweep. Because the wavelength of the beeper (4 inches) is small compared to the diameter of the opening (12 inches), very little diffraction should occur, and listeners should hear a much louder sound during that part of the sweep when the carpet cylinder is pointing directly toward them. Now try the same thing with the beeper placed in the 4-in.-diameter carpet cylinder. Since now $\lambda/d = 1$, the diffraction should be appreciable, and the sound should show much less change in intensity as the direction of the carpet cylinder is changed.

In addition to showing the effect of varying $d$, you can also show the effect of varying $\lambda$. Take both end caps off, and talk through each carpet cylinder in turn as you sweep its direction relative to the listeners. In contrast to the results you obtained using the high-frequency beeper, *both* carpet cylinders should show significant diffraction, since the average wavelength of your voice probably exceeds twelve inches. For best results, talk in a deep voice if your natural voice is very high pitched.

You could produce a carpet cylinder that produces even less diffraction than the 12-in.-diameter one, but this would require a larger diameter and a *much* larger length. The length is important, because in addition to the spreading out of sound due to diffraction, a geometrical spreading also occurs. For a point source a distance $L$ in front of an opening of diameter $d$, the geometric spreading angle is $\tan^{-1}(d/2L)$. The total spreading angle can be approximated by a quadratic combination of the geometric spreading angle and the diffraction spreading angle $\sin^{-1}(1.22\lambda/d)$. The table below shows the rolled-up carpet cylinder dimensions (length and diameter) to use to achieve a given spreading angle. Ideally, the sound should drop to a very small fraction of its forward intensity at the listed angles. The values of $d$ and $L$ are tabulated as multiples of the wavelength $\lambda$.

| Total spreading angle (degrees) | Diameter (in wavelengths) | Length (in wavelengths) |
|---|---|---|
| 60° | 1.80 | 0.98 |
| 45° | 2.31 | 1.86 |
| 30° | 3.36 | 4.32 |
| 15° | 6.63 | 17.7 |

According to the preceding table, to achieve a total spreading angle of 15° using a source with a 4-in. wavelength, you must use a carpet cylinder with a diameter of 6.63 wavelengths (26.5 in.), and a length of 17.7 wavelengths (70.8 in.). Such a cylinder would be somewhat unwieldy to sweep across an audience, although you could have the carpet fixed and the listeners pass in front of it. You could, however, achieve a small spreading angle without such a large cylinder by using a much shorter sound wavelength. For example, most so-called "silent" dog whistles that are manufactured in America produce a sound that is near the upper end of the audible range for the average listener.

An alternative version of this demonstration, which involves much less construction, is to replace the small carpet cylinder by a mailing tube, and the large cylinder by a 12-in.-diameter plastic bottle from a water cooler. If you cut the flat end of the bottle off, you can use it as a megaphone, and either talk through it (low frequencies) or blow a whistle through it (high frequencies) while sweeping it across an audience. The variation in sound intensity with angle is, however, not quite as pronounced as it is when you use the carpet cylinder, because carpeting is a better sound absorber and prevents multiple reflections of the sound as it passes down the cylinder.

Still another variation on the demonstration would be to use a large plastic trash container of rectangular cross section (suggested size 9 in. × 18 in. × 36 in.). If you cut a hole in the bottom of the container, you can whistle through the hole while sweeping the container across the audience. Listeners should observe more sound directionality (less diffraction) when the long dimension of the rectangular opening is horizontal than when it is vertical.

## Q.6. Moving a beeper below desk level

### Demonstration
The extent to which diffraction around an obstacle depends on wavelength is shown when sound sources that have different wavelengths slowly descend behind an obstacle such as a desk.

### Equipment
The 12-in.-diameter carpet cylinder and the Radio Shack beeper that were described in demonstration Q.5.

### Comment
Use the carpet cylinder with the beeper for this demonstration, to make the sound waves that are incident on the back of the desk fairly directional—i.e., plane waves rather than spherical waves. As the beeper gradually descends below the level of the desk, listeners should hear a noticeable drop in sound intensity, because the high-frequency sound from the beeper does not diffract over the edge of the desk very much. (Be sure the rolled-up carpet is pointing horizontally.) Now you can show that low-frequency sound diffracts to a much greater extent, by ducking down below the desk level while you are talking. Listeners will hear hardly any change in the loudness of your voice.

## Q.7. Beeper at the center of a hollow tube

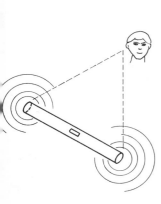

### Demonstration
A two-source interference experiment can be performed using a hollow tube with a single beeper placed at its center.

### Equipment
A Radio Shack beeper (see demonstration J.5); and a hollow tube about two feet long and 2 to 3 inches in diameter (a mailing tube, for example). The tube's diameter has to be larger than two inches so that the beeper will fit into it, and the tube should be reasonably sound absorbent.

### Construction/Comment

Hold the hollow tube vertically, and lower the beeper and its battery—connected to a wire—into the tube until the beeper is at the tube's center. Activate the beeper by connecting the ends of the wire together, and hold the tube horizontally. The beeper at the center of the tube produces sound waves that travel mirror-image paths to reach each end; the effect is like having two coherent sound sources located at the ends of the tube. Since the tube's diameter is small compared to the beeper's wavelength, the sound coming out of each end will radiate in all directions. (When you lower the beeper down the tube, make sure the front of the beeper faces into the wall of the tube and not toward one end, to ensure that the sound from each end will have the same intensity.)

Rotate the tube in a horizontal plane through a range of positive and negative angles. ($A = 0°$ corresponds to the tube being at right angles to a line from the center of the tube to the listener.) As the tube is rotated, the listener should hear fluctuations in sound intensity corresponding to the maxima and minima in the interference pattern. (Of course, no such fluctuations in sound will be heard if the tube is kept stationary.) The angles along which maxima in the interference pattern will be heard are given by the condition $\sin A = m\lambda/d$. For $\lambda = 4$ inches, and a tube length $d$ of 24 inches, this equation yields $\sin A = 0$, $\pm0.167$, $\pm0.333$, $\pm0.500$, $\pm0.667$, $\pm0.833$, and $\pm1.000$, a total of 13 maxima between $A = +90°$ and $A = -90°$. In verifying this result, it may be easier to count the number of maxima than to determine the exact angles at which they occur. Don't expect exact agreement, however, since the beeper wavelength may differ from its nominal value of 4 inches.

## Q.8. Acoustic diffraction grating

### Demonstration

An acoustic diffraction grating, made from a pipe with holes in it and a beeper at one end, can be rotated to produce maxima and minima heard by a distant listener.

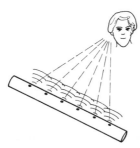

### Equipment

A Radio Shack beeper (see demonstration J.5); a 4-ft-long piece of 2-in.-diameter polyvinyl chloride (PVC) pipe; two end caps for the PVC pipe; and small pieces of carpeting or other sound-absorbing material.

## Construction

Drill a row of five ¼-in.-diameter holes, 8 inches apart, on the PVC pipe. Also drill a ⅜-in.-diameter hole near one end of the PVC pipe, on the side opposite the row of five holes. Put carpeting or other sound-absorbing material in the end caps. Insert the beeper snugly into the end of the PVC pipe, with its wire passing out through the ⅜-in.-diameter hole. Seal the PVC pipe at both ends with the end caps, but don't cement them on.

## Comment

Hold the PVC pipe with the row of holes facing a listener. Activate the beeper, and slowly rotate the pipe through a large range of angles. The listener should hear the sound intensity vary as the pipe is rotated. It may help if the listener covers one ear, and orients the other ear in the direction of the sound. It is also easier to hear distinct maxima if the listener is far away. If the listener is too close, there won't be a constant phase difference from hole to hole because the path length to the listener increases nonuniformly from hole to hole.

Ideally, the minimum distance from the PVC pipe to the listener should be $d^2/\lambda$, where $d$ is the distance between holes 1 and 5. At this distance, waves reaching the listener from the two end holes have an extra phase difference of $\pi/4$ relative to the waves from the middle hole. For $d = 4$ feet and $\lambda = 4$ inches, the suggested minimum distance to the listener works out to be 48 feet, which may be difficult to achieve indoors. Remember that when the listener is far away, little sound intensity will be heard except at those angles at which principal maxima occur. If you are not able to achieve a 48-ft source-to-listener separation, the listener will still hear clear variations in sound intensity as the pipe is rotated, but the pattern will be more complex.

## Q.9. Squinting at an unfrosted bulb

### Demonstration

Squinting while you look at the filament of an unfrosted light bulb allows you to see a diffraction pattern.

### Equipment

An unfrosted 60-watt light bulb, preferably one with a linear filament.

**Comment**

If the filament in the light bulb is vertical (as most are), turn the light 90° so that the filament's smaller dimension is vertical. This makes the dimensions of the source as small as possible in the direction perpendicular to the slits formed by your eyelids. Station yourself some distance from the light (at least 10–15 feet away) in a darkened room, and squint one eye, keeping the other eye closed. Experiment until you see a pattern like that shown in the illustration. The short equally spaced horizontal lines are the diffraction maxima formed when light passes through the narrow slit between your eyelids. (However, see the discussion on "Mach bands" in the following demonstration.)

## Q.10. Variable-width slit using razor blades

**Demonstration**

Single-slit diffraction patterns can be produced using a variable-width slit made from two razor blades, and the inverse dependence of the extent of diffraction on the slit width can be shown.

**Equipment**

Two razor blades; a piece of transparent material (glass, plastic, or film); and an unfrosted light bulb—preferably one with a linear filament.

**Comment**

Tape the two razor blades onto the transparent backing, with their edges in contact at one end, and separated by a fraction of a millimeter at the other end, thus forming a slit of variable width. Hold the slit up, and look through it at a light background. You should see a pattern of vertical lines in the space between the slits. These lines are so-called "Mach bands"—an artifact of the visual system, not a product of light diffraction.

Now take the slit into a darkened room, and look through it at an unfrosted light bulb placed 10 to 20 feet away. After some careful observation, you should see a diffractionlike pattern whose width increases when you look through the portions of the slit that have smaller width. In other words, dark lines (the minima) will be seen to curve away from the central maximum in the direction of decreasing slit width. The dark lines show the greatest curvature when you look at the source through the end of the slit (where the width $d$

approaches zero), because the angular width of the diffraction pattern varies inversely with $d$, and so the dark lines have a hyperbolic shape. The curvature of the dark lines shows that the effect is due to diffraction, and not to Mach bands (which would be parallel to the slit edges).

It may seem surprising that the pattern appears essentially monochromatic, because the different wavelengths that are present in white light should diffract by different amounts, resulting in dispersion. It is only for a large number of closely spaced parallel slits, however, that appreciable dispersion is observed, owing to the narrowness of the principal maxima in this case (see demonstration Q.11). You can readily confirm this if you have a test pattern comprised of different numbers of slits through which you can view the white-light source. The degree of dispersion (the amount of color seen) gradually decreases as you go from many closely spaced slits to many slits with wider spacing, and finally to two- and three-slit patterns, where the much broader maxima make dispersion hardly noticeable.

## Q.11. Diffraction-grating sheets

**Demonstration**
Using pieces of plastic diffraction-grating sheet, you can observe the spectra of various sources and measure their wavelength(s).

**Equipment**
A large sheet of plastic diffraction-grating material (from Edmund Scientific Company, for example); 35 mm slide frames (not necessary, but convenient, if you wish to reuse the diffraction-grating squares); and a variety of light sources. It would also be worthwhile to include reflecting diffraction foil in the demonstration.

**Construction**
Cut the diffraction-grating sheets into small squares that can be taped onto the 35 mm slide frames. (You could also tape them onto 3×5-in. index cards with holes to look through.)

**Comment**
Inexpensive diffraction-grating sheets provide a good way to demonstrate diffraction of light to a large group. Each member of the group can look through the diffraction gratings at a variety of sources. Particularly good sources would

be an unfrosted light bulb, a neon lamp, and some heated gas sources (for discrete spectra). Try rotating the diffraction grating while you look through it, holding it close to your eye. If you know the spacing $d$ between the lines of the grating, you can compute the wavelength(s) of the light you are looking at by estimating the angle to the first-order spectrum $A$, and calculating $\lambda$ from $\lambda = d \sin A$. (The angle $A$ can be estimated from the equation $A = \tan^{-1}(y/D)$, where $y$ is the linear displacement of the first-order spectrum and $D$ is the distance between you and the source. It may be helpful to have a meterstick right behind the source to help estimate $y$. Of course if you are looking at a white light, you will be able to compute only the average wavelength of the spectrum.

Observe what happens when you rotate the grating by 90°, and also what happens when you look through two diffraction grating sheets whose lines make a 90° angle. One type of reflecting diffraction foil, which consists of a pattern of many small circular reflection gratings, also demonstrates diffraction of light very nicely. One advantage of the reflection foil is that each viewer need not have a separate grating—all observers can view the foil when it is placed behind the light source. Diffraction patterns can also be seen outdoors when street lights are viewed through umbrellas.

## Q.12. Pin holes in aluminum foil

### Demonstration
Diffraction and interference patterns can be seen when you look at an unfrosted light bulb through pinholes in aluminum foil.

### Equipment
Aluminum foil; a fine needle; and an unfrosted light bulb, preferably one with a small filament.

### Comment
Make a number of holes in the aluminum foil using the needle; keep them as small as possible. (It may help to put the aluminum foil on a hard surface before you stick the needle into it.) In one region of the foil, create pairs of holes, with the spacing between the two holes in each pair as small as possible. When you darken the room and view the unfrosted light bulb through the single holes from 10–15 feet away, you should see diffraction rings. It may be difficult to see very many concentric rings, but you should certainly see a

central bright region surrounded by a dark ring and a bright one. If the holes are not exactly circular, or if the size of the light bulb filament is appreciable, the regularity of the rings will be destroyed, but some more complex and unusual patterns may be produced. When you view the light bulb through the double holes, you should see a pattern of stripes—interference fringes—superimposed on the diffraction pattern. The spacing of the stripes is inversely proportional to the distance between the two adjacent holes, so the holes must be quite close together to get noticeable stripes.

## Q.13. Hologram eyeglasses

### Demonstration
Gag eyeglasses with holographic images of eyeballs are a nice illustration of holography.

### Equipment
Hologram eyeglasses, obtainable from Toys-R-Us for about $3.00. Reflection holograms can also be obtained for about $20.00 (for a 5×7-in. size) from the Smithsonian Institution museum gift shop in Washington, D.C.

### Comment
A hologram results from the interference between light reflected off an object and light from the illuminating source. The interference pattern recorded on very fine grain photographic film constitutes a complete record of the light wave reflected off the object. In contrast to a normal photograph, which records only light intensity at each point, the hologram—because it uses interference—also records the phase of the light at each point. It is necessary to record the hologram using a vibration-free setup and very fine grain film because the interference fringes on film are very closely spaced—too close to be seen by the naked eye. When the hologram is illuminated at the proper angle, the original reflected wave is re-created, and the object is seen as three dimensional.

## Q.14. Interference in thin films

### Demonstration
The pattern of stripes (fringes) associated with interference in a thin film of varying thickness can be observed using a soap film.

### Equipment

Soap-bubble solution; a circular soap-bubble frame; a large convex lens (or Fresnel lens); mounting stands for the lens and soap-bubble frame; and a focusable flashlight or slide projector. A good soap-bubble frame can be made using a tin can with a small hole drilled in the bottom. The can shields the bubble film from breezes, and the small hole in the bottom prevents a pressure differential between inside and outside. You can attach a threaded $\frac{1}{8}$-in.-diameter rod to the middle of the can, which will make it easier to mount the can on a vertical stand after dipping the rim of the can into soap-bubble solution.

### Comment

Light reflecting off the front and back surfaces of a thin soap film interferes constructively or destructively, depending on the film thickness at any given point. If the thickness is much less than one wavelength, the interference is destructive, owing to a 180° phase reversal for the light reflecting off the front surface. The interference is also destructive for normal incidence if the thickness is an even multiple of a quarter of a wavelength, and it is constructive for odd multiples. Thus at any given point on the film, some wavelengths interfere constructively and others destructively, which accounts for the beautiful colors seen.

To demonstrate interference in a thin soap film, dip the frame into a soap-bubble solution and set it on a vertical stand. Darken the room, and reflect the light from a focusable flashlight or slide projector off the soap film. Use the lens to focus the image on a screen. Initially, the image of the soap film appears irregular, but a pattern of colored horizontal stripes—associated with a film of uniformly varying thickness—soon begins to emerge. As the soap solution drains, and the film gets thinner (and its thickness more uniform), the spacing between the stripes increases. Just before the film breaks, a black stripe forms at the top of the film (the bottom of the inverted image). The film appears black at this point because the thickness is much less than one wavelength, and the 180° phase difference for all wavelengths results in destructive interference for all colors.

## Q.15. Diffraction rings through a mist

### Demonstration

You see diffraction rings when you view a bright light through a fine mist on a glass onto which you have exhaled.

**Interference & Diffraction: Optical**

The average size of the water droplets can be determined from the size of the rings.

**Equipment**
An unfrosted light bulb; a meterstick; and a small piece of glass (or a pair of eyeglasses, if you wear them).

**Comment**
Diffraction rings, also called coronas, are seen in a variety of natural situations, including viewing the moon through thin clouds, or viewing street lights through a fog. The diffraction results when light waves bend around fine water droplets by an amount that depends on the size of the droplets. In order to see a distinct ring when you view a light source through many droplets, the droplets must all have about the same diameter—otherwise, the rings from different droplets would have different radii and the pattern would wash out.

It is surprisingly easy to duplicate the diffraction rings seen in nature by exhaling and fogging a piece of glass or your eyeglasses, causing a fine mist to be deposited. If you view an unfrosted light bulb through the glass in a darkened room, faint colored rings surrounding a central blob can be seen for a few seconds before the mist evaporates. (If the glass is chilled before fogging, the condensed mist persists much longer.) The central blob will have blue on the inside and red on the outside, because shorter wavelengths exhibit less diffraction. You probably will not see more than one faint ring around the central blob, unless the source is very bright.

Hold a meterstick at arm's length, and estimate the radius to the edge of the central blob, $r$. Find the angular radius in radians by dividing $r$ by the length of your arm. The angular radius $A$ of the edge of the central blob (the first minimum in the diffraction pattern) is related to the wavelength $\lambda$, and to the diameter $d$ of the water droplets, through the equation $\sin A = 1.22\lambda/d$. Thus you can calculate the size of the water droplets causing the diffraction using the measured angular radius $A$, and a value of $\lambda = 600$ nm for the average wavelength of light. You will probably observe an angular radius of a few degrees, which corresponds to a droplet diameter of 0.02 mm—about the same size as some cloud droplets.

# Appendix 1

---

## Other Books of Physics Demonstrations

A good summary of the literature on physics demonstrations up to 1979 can be found in John A. Davis and Bruce G. Eaton, "Resource Letter PhD-1: Physics Demonstrations," *American Journal of Physics 47*(1), 835–40 (October 1979). Some well-known books include:

Blasi, Rocco C., ed. *Physics Fun and Demonstrations with Professor Julius Sumner Miller*. Franklin Park, Ill.: Central Scientific Company, 1974.

Edge, R. D. *String and Sticky Tape Experiments*. College Park, Md.: American Association of Physics Teachers, 1981.

Freier, G. D., and F. J. Anderson. *A Demonstration Handbook for Physics*. College Park, Md.: American Association of Physics Teachers, 1972.

Meiners, H., ed. *Physics Demonstration Experiments*. New York: Ronald Press, 1970.

Sutton, R. *Demonstration Experiments in Physics*. New York: McGraw-Hill, 1938.

VanCleave, Janice Pratt. *Teaching the Fun of Physics*. New York: Prentice Hall, 1985.

# Appendix 2

## Equipment Suppliers Cited in This Book

Andrews Manufacturing Company, Inc.
2121 Franklin Blvd.
Eugene, OR 97403

Central Scientific Company
11222 Melrose Avenue
Franklin Park, IL 60131

Edmund Scientific Company
101 East Gloucester Pike
Barrington, NJ 08007

Index Packaging, Inc.
Rte. 16, Box 367B
Union, NH 03887

Jerryco, Inc.
601 Linden Place
Evanston, IL 60202

Nakamura Scientific Company
U.S. Distributor:
NADA Scientific Ltd.
P.O. Box 1336
Champlain, NY 12919

RunTronics, Inc.
1301 Shoreway Road
Belmont, CA 94002
(Demonstration A.5 suggests a substitute device for the RunTronics product.)

Toltoy, Inc.
6444 Telegraph Road
Erie, MI 48133

Ward's Natural Science Establishment, Inc.
5100 West Henrietta Road
P.O. Box 92212
Rochester, NY 14692

# Index